AN ANTHOLOGY
OF EXQUISITE

Birds

Written by Ben Hoare

Illustrated by Angela Rizza
and Daniel Long

Introduction

When I was a young boy, an unusual pink, white, and black bird flew in front of me. Later, I learned it was a hoopoe, and seeing it started my life-long passion for birds. After reading this book, I hope you will love them as much as I do! Birds are brilliant. They are masters of the air and the only animals alive today with feathers. In fact, the first animals with feathers were dinosaurs, from which birds are descended. And that means every single bird is a living dinosaur, which is pretty mind-boggling!

Birds lead fascinating lives, full of drama. They eat just about everything imaginable, sing beautifully, build fabulous nests, make friends, dance acrobatically, dive deep underwater, play tricks on each other, and even use tools. Best of all, birds are everywhere — in city centres, forests, and deserts, at the top of mountains, and far out at sea. This book is a celebration of all these exquisite feathered creatures, which we are so lucky to share the Earth with.

Ben Hoare
Author

Contents

Green peafowl 4	Hoatzin 76	Blue jay 148
Dalmatian pelican 6	Resplendent quetzal 78	Bearded bellbird 150
Emu 8	Kākāpō 80	Namaqua sandgrouse 152
Marabou stork 10	Helmeted guineafowl 82	Cuban trogon 154
Secretary bird 12	Toco toucan 84	Killdeer 156
Red-crowned crane 14	Greater roadrunner 86	Flame bowerbird 158
Black swan 16	Harlequin duck 88	Long-tailed manakin 160
Wandering albatross 18	Great crested grebe 90	European starling 162
Andean condor 20	Cuckoo-roller 92	*Wings* 164
Bearded vulture 22	Montezuma oropendola 94	Northern cardinal 166
Bird evolution 24	*Nests* 96	Noisy pitta 168
Harpy eagle 26	Oilbird 98	White-throated dipper 170
Great bustard 28	Sunbittern 100	Bohemian waxwing 172
Red-tailed tropicbird 30	Green heron 102	Greater honeyguide 174
Hyacinth macaw 32	Laughing kookaburra 104	Budgerigar 176
Great frigatebird 34	Northern potoo 106	Common poorwill 178
Lesser flamingo 36	Pileated woodpecker 108	Little bee-eater 180
Beaks 38	Little spotted kiwi 110	Common swift 182
Superb lyrebird 40	Little penguin 112	*Feet* 184
Great cormorant 42	Peregrine falcon 114	Wilson's bird of paradise 186
Great northern diver 44	Cape sugarbird 116	House sparrow 188
Red-legged seriema 46	Livingstone's turaco 118	Woodpecker finch 190
Rhinoceros hornbill 48	Bar-tailed godwit 120	Sociable weaver 192
Pheasants 50	Snow petrel 122	Long-tailed tit 194
Anhinga 52	Pigeons 124	European robin 196
Roseate spoonbill 54	Grey parrot 126	Elf owl 198
Blue-footed booby 56	Tufted puffin 128	'I'iwi 200
Snow goose 58	Scissor-tailed flycatcher 130	Superb fairywren 202
Great bittern 60	Eurasian coot 132	Black-backed kingfisher 204
Red junglefowl 62	Arctic tern 134	Common tailorbird 206
Australian brushturkey 64	Common cuckoo 136	Cuban tody 208
Scarlet ibis 66	*Eggs* 138	Hummingbirds 210
Great grey owl 68	White tern 140	Tree of Life 212
Feathers 70	Ruff 142	Glossary 214
Common raven 72	Eurasian hoopoe 144	Visual index 216
European herring gull 74	Andean cock-of-the-rock 146	Acknowledgements 224

Green peafowl

Peafowl feathers glitter when sunlight catches them.

Green peafowl, Southeast Asia. Male peafowl, known as peacocks, strut about with their tails spread when displaying to watching females, called peahens.

"Awow—woh!" What could that bizarre sound be? It's the cry of the male green peafowl, of course! In the morning and evening, he calls out to tell other peafowl that he's looking for a mate. If a female peafowl hears his cry, she answers "ow—aah!" and walks through the forest to find him.

The male peafowl dances to try and impress the female. First, he opens his long tail into a fan to show off the eye-like spots on the feathers. Then, he shakes his tail, making the spots sparkle. The more spots he has, the more likely the female is to be interested! There is another kind of peafowl in Asia, the Indian peafowl, which has a beautiful metallic blue body.

Pelican chicks stick their head inside their parent's beak to get a fishy meal.

Dalmatian pelican

A pelican can turn its colossal beak into a brilliant fishing net. The beak has a pouch of stretchy skin, which the pelican uses to scoop up fish. But the pouch fills up with water as well as fish! So the pelican has to open its beak a little to let the water drain out, before gulping down its dinner.

Dalmatian pelicans often swim in a line to herd fish and make them easier to catch. They are huge birds that weigh as much as a small child. To take off from the lakes and marshes where they live, they need to run across the water with their massive wings flapping. Once in the sky, they soar like birds of prey, sometimes as high as planes.

Dalmatian pelican, Asia and Europe. A pelican's beak holds 11 litres (2.5 gal) of water, which is more than its stomach can.

In emus, it is the devoted dads which look after the young.

Emu

This is not a dinosaur foot. In fact, it belongs to the emu. Everything about the emu is HUGE. It is the second tallest bird on the planet, after the ostrich; it weighs as much as an average 12-year-old human; and its footprints are whoppers — they're almost as long as the height of this book!

However, the emu's wings are so tiny, you can hardly see them. And its fluffy feathers are no good for flying, either. Instead, the flightless emu walks everywhere. To escape danger, it runs. The emu would easily beat the world's fastest human in a race. If this powerful bird has to face an enemy, it can give an almighty kick to defend itself.

Emu, Australia. The emu's colossal feet have three strong toes, each armed with a sharp claw.

Marabou stork

Marabou storks are some of the biggest flying birds. They soar high above the grasslands of Africa to look for food. Mostly the storks feed on the bodies of dead animals, but they kill prey too. Their massive beak tears through skin and fur to reach the meat inside. You may think that their bare head looks strange, however, it allows them to eat without getting their feathers dirty.

And what is that odd pink cone under the stork's beak? It's a saggy bag of skin — and the stork can inflate it like a balloon! It does this during squabbles and when showing off to a breeding partner. Some people think marabou storks are ugly or scary, but really they're wonderful, wouldn't you agree?

Marabou storks are drawn to lion kills, where they eat the leftover meat.

Marabou stork, Africa. Both male and female marabou storks have a pouch of skin underneath their throat.

Secretary bird, Africa. Secretary birds are one of the few birds that have eyelashes.

Secretary bird

Many birds of prey swoop down or glide to catch their dinner. Secretary birds are different! These tall birds stride across the plains of Africa and hunt on the ground. They eat rats, snakes, and other small animals, which they stamp on with their long, strong legs. Whenever there is a fire, secretary birds will run towards the flames to catch other animals that are trying to escape. But these unusual birds can still fly when they need to.

How did secretary birds get their name? No one is sure, but it may be to do with their black crest. You see, their crest feathers look like the feather quills that secretaries used as pens many years ago.

Secretary birds bang snakes on the ground to kill them.

Red-crowned crane

It seems these tall, elegant birds can't stay still for long. Whenever the cranes meet, they soon start dancing. They bow to one another, and jump into the air with a quick flap of their wings. Sometimes they even play chase, like a group of children! In their courtship dance, the male and female pair dance gracefully together.

Red-crowned cranes live in marshes, rice fields, and other wet places. They are much-loved birds in Japan and China, where they are seen as a sign of good luck and a long life. Northern Japan has cold, snowy winters and people put out rice and corn for the cranes, which is helping the crane population to grow.

In Japan, these birds are known as "gods of the wetland".

Red-crowned crane, eastern Asia. Dancing cranes copy their partner's moves to perfection, raising their wings and stretching their necks.

Black swan

Most swans have mainly white feathers, but black swans are as dark as the night sky. You can see these handsome birds on lakes and ponds all over Australia, even in busy cities. Often the swans will be in small groups. Each group is a family that includes a breeding pair and their young, called cygnets. It's easy to tell the youngsters apart, because they are grey, not black.

The swans build a very messy nest. It's a humongous heap of water weeds, grass, and anything else that they can find. Both parents defend their nest fiercely. If a predator comes near, they hiss and chase it away with their wings flapping furiously.

Black swans make a sound like someone blowing a plastic toy trumpet!

Black swan, Australia.
These swans hold their neck in a graceful curve as they paddle through the water.

Wandering albatross

The wandering albatross may live for 50 years or more. Most of that time, it is alone, soaring over the ocean, out of sight of land. It is huge — as heavy as a goose — with the longest wings of any bird. Thanks to these mighty wings, it can ride the ocean wind and travel enormous distances with hardly a flap. It has a special beak with two tube-like nostrils on top. These help it smell food, mainly fish and squid, from up to 20 km (12 miles) away.

Albatrosses stay with the same partner for life. After long periods apart, the pair greet each other on their nesting island by clattering their beaks together. They are devoted parents, and spend nearly a year raising each chick.

The wingspan of these magnificent birds is almost 3.5 m (11.5 ft) across.

Wandering albatross, Atlantic, Pacific, and Southern Oceans. These albatrosses breed on remote islands and rear one chick at a time.

Andean condor

Condors ride currents of rising air to lift them into the sky.

Andean condor, South America. A condor's huge wingtip feathers give it extra control as it flies.

The Andean condor is a gigantic bird with a strange, bald head. It weighs as much as a swan and its wings are some of the largest in the world. Most of these condors live in the Andes Mountains in South America, where they soar high over towering peaks. Amazingly, the birds can glide for hours without a single flap of their mighty wings.

Condors are a kind of vulture and feed on dead animals. Since their sense of smell is not great, they rely on superb eyesight to spot food. A condor can wolf down so much meat in one go, it's the equivalent of eating a small dog. Sometimes, the bird is even too full to take off!

A bearded vulture's diet consists almost entirely of bones.

Bearded vulture

Vultures are huge birds with hooked beaks and sharp claws. Most of them feed on the remains of dead animals and do not kill prey. However, bearded vultures ignore the meat and eat the bones instead. They swallow small bones whole, but carry large ones into the air and drop them on rocks to smash them to pieces. Their stomach is full of super-strong acid that dissolves the bones quickly.

Bearded vultures are named for the tuft of black feathers under their beak, which looks like a beard. You can see these birds soaring high over mountains. They like to bathe in pools where iron in the water turns their chest feathers orange. No one knows why!

Bearded vulture, Africa, Asia, and Europe. Bearded vultures have sharp eyesight to spot food far away.

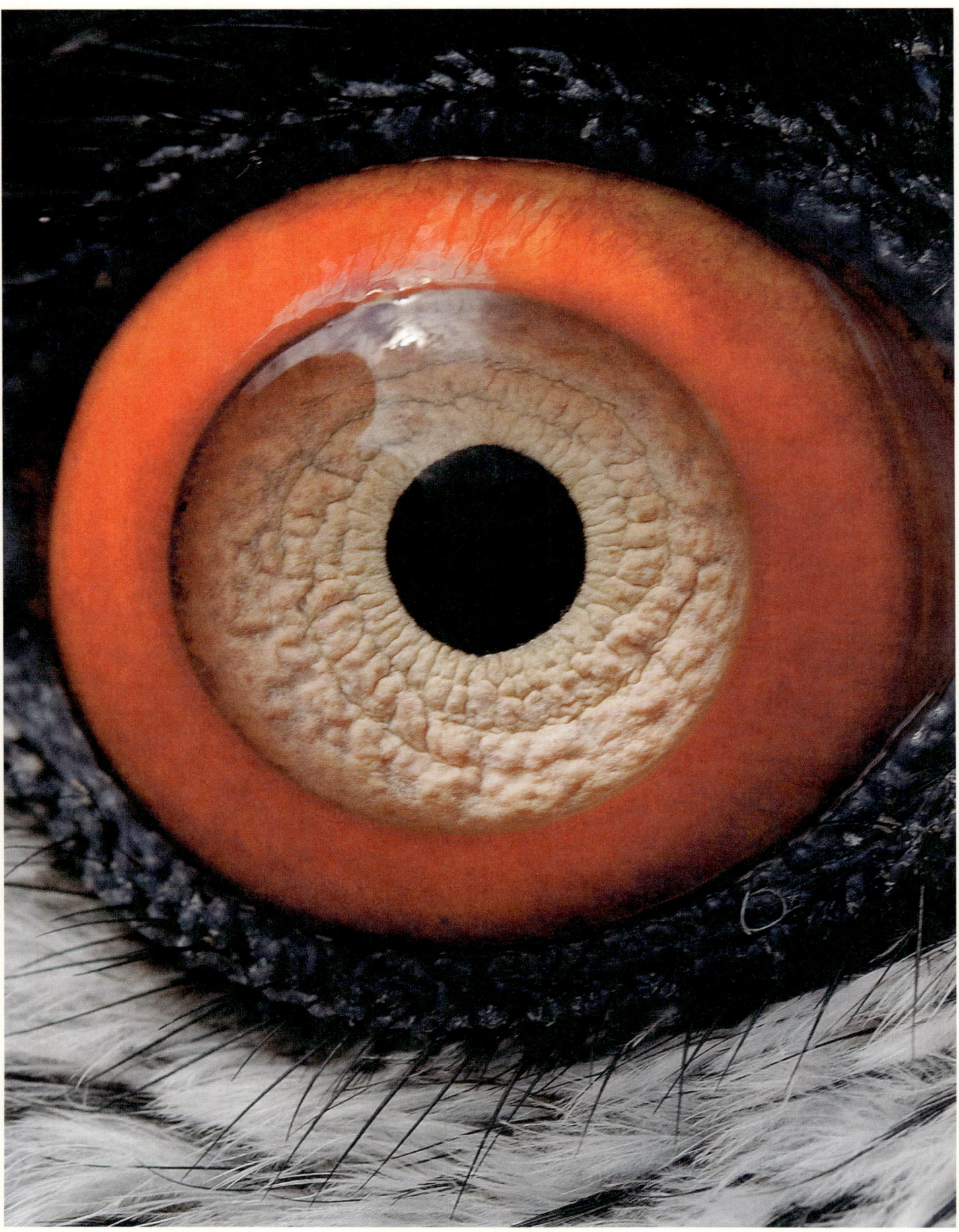

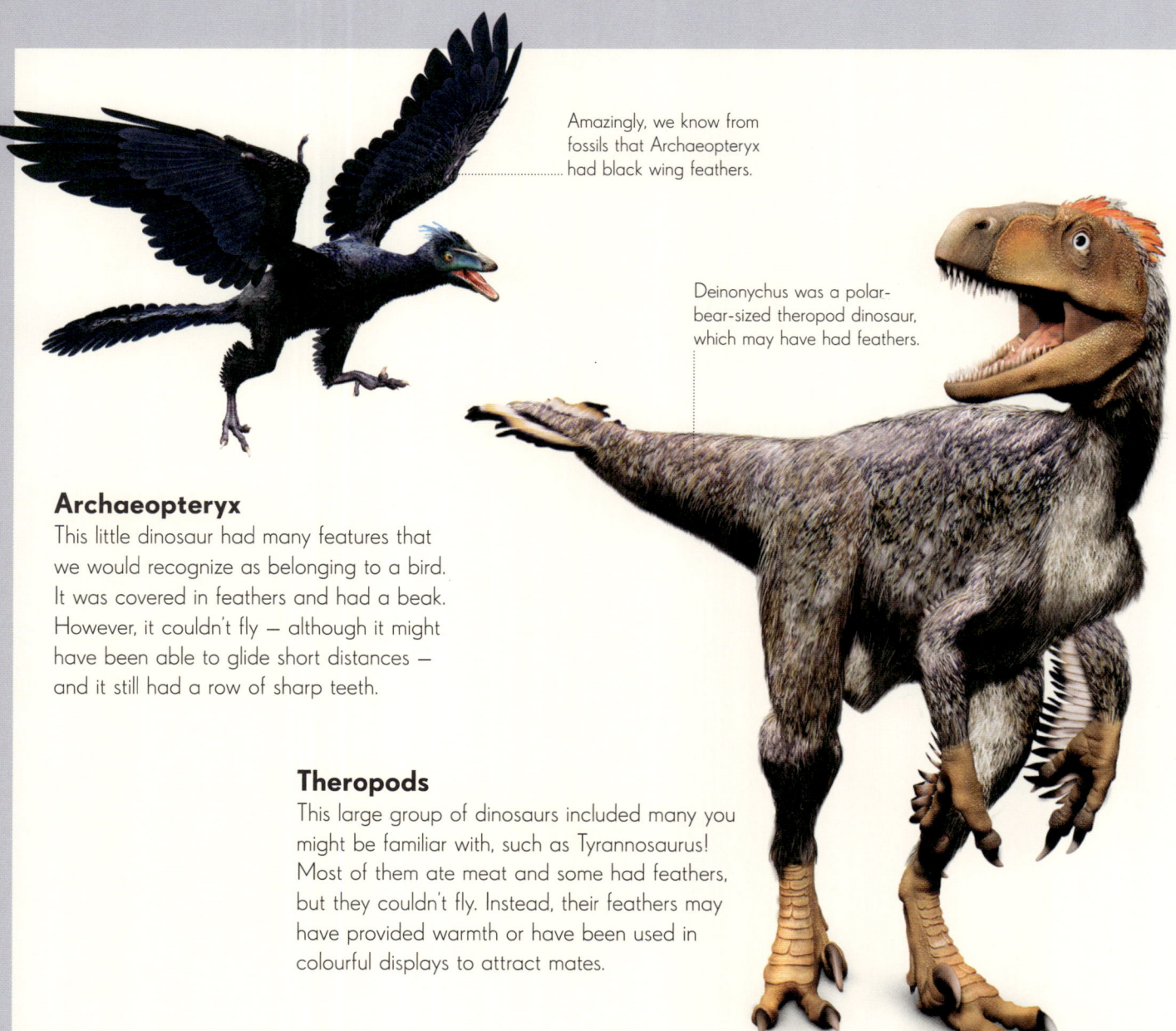

Amazingly, we know from fossils that Archaeopteryx had black wing feathers.

Deinonychus was a polar-bear-sized theropod dinosaur, which may have had feathers.

Archaeopteryx
This little dinosaur had many features that we would recognize as belonging to a bird. It was covered in feathers and had a beak. However, it couldn't fly — although it might have been able to glide short distances — and it still had a row of sharp teeth.

Theropods
This large group of dinosaurs included many you might be familiar with, such as Tyrannosaurus! Most of them ate meat and some had feathers, but they couldn't fly. Instead, their feathers may have provided warmth or have been used in colourful displays to attract mates.

Bird evolution

Today, birds live in every part of the world from pole to pole. Where did they come from though? It might surprise you to know that birds are, in fact, dinosaurs! About 95 million years ago the first birds evolved from a group of dinosaurs called theropods, some of which, although they were reptiles, had feathers. Over many generations these early birds lost their teeth and tail, developed beaks, and even took to the skies! Now, there are more than 11,000 species of birds in the world.

Confuciusornis
This early bird was about the size of a crow. Its fossils show it had feathers and wings, which suggest it could fly short distances. It also had a much shorter tail than other early birds, although it had two long tail feathers.

Like other early birds, Confuciusornis had claws on its wings.

Ichthyornis
Ichthyornis looked a bit like seabirds do today, and lived along coasts. It had a strong breastbone for flight muscles to attach to and was a good flier. However, it still had teeth in its beak, which it probably used to catch fish.

Asteriornis
Asteriornis is the earliest bird yet discovered that shares many features with living birds. It belonged to the same group as chickens — and its nickname is the "Wonderchicken"! It lived at a time just before non-bird dinosaurs went extinct, and had a toothless beak. It almost certainly had feathers, and would probably have been able to fly.

Asteriornis had long legs and found food on the ground on the coast.

Living birds
There is now a huge number of different types of bird of all shapes and sizes. All have a toothless beak and feathers, and most can fly. Take a close look though, and you can still see some of the features of their ancient dinosaur relatives.

Harpy eagle

A harpy eagle's talons are as long as the claws of a brown bear.

Harpy eagle, Central and South America.
A harpy eagle has a long talon at the end of each of its four toes. Three point forwards and one points backwards.

Monkeys dive for cover when they spot a harpy eagle. This bird is one of the largest eagles in the world, and monkeys are on its menu. The mighty eagle grabs its prey straight from the treetops with its powerful talons. Sloths are probably this eagle's favourite food. Despite its great size, it flies silently and takes the sloths by surprise as they snooze in the trees.

The harpy eagle lives in rainforests, and a pair will stay together for life. Together, they build a massive nest the size of a double bed that would be strong enough for you to stand on! The eagles raise one chick at a time and feed their huge youngster for as long as 10 months.

Great bustard

This bird might remind you of a massive chicken, but unlike a chicken, it can fly well. The male great bustard is one of the heaviest of all flying birds. Any heavier, and his wings wouldn't lift him into the air! Most of the time, the bustard prefers to walk.

Great bustards roam grasslands and fields. They gobble up most edible things they come across, including seeds as well as animals such as insects, mice, and frogs. In spring, the male bustards gather together to show off to females. During their group display, the bustards puff out their neck and spread their tail and wing feathers. This turns them into big white balls of fluff!

Male great bustards grow a moustache of whiskery feathers.

Great bustard, Asia and Europe. Male great bustards look very similar to females, except for when they are displaying.

Red-tailed tropicbird

A white speck is moving far out at sea. As you watch it come closer, you realize it's a bird... with an amazing tail! This is a red-tailed tropicbird. The two red feathers in the middle of its tail are so long, they trail behind like the streamers on a kite.

Tropicbirds spend most of their life flying over the waves. They dive to catch fish and squid to eat. Sometimes they rest on the ocean, but despite having webbed feet like gulls and ducks, tropicbirds are not great swimmers. Only one thing makes these seabirds visit land — the need to raise a family. Usually they nest on remote islands, far away from any people.

Red-tailed tropicbird, Indian and Pacific Oceans. The long, red feathers of a tropicbird's tail are sometimes longer than its body.

Tropicbirds make a scruffy nest, often just a scrape in the ground.

Hyacinth macaw

Hyacinth macaws will fly a long way to find their favourite nuts.

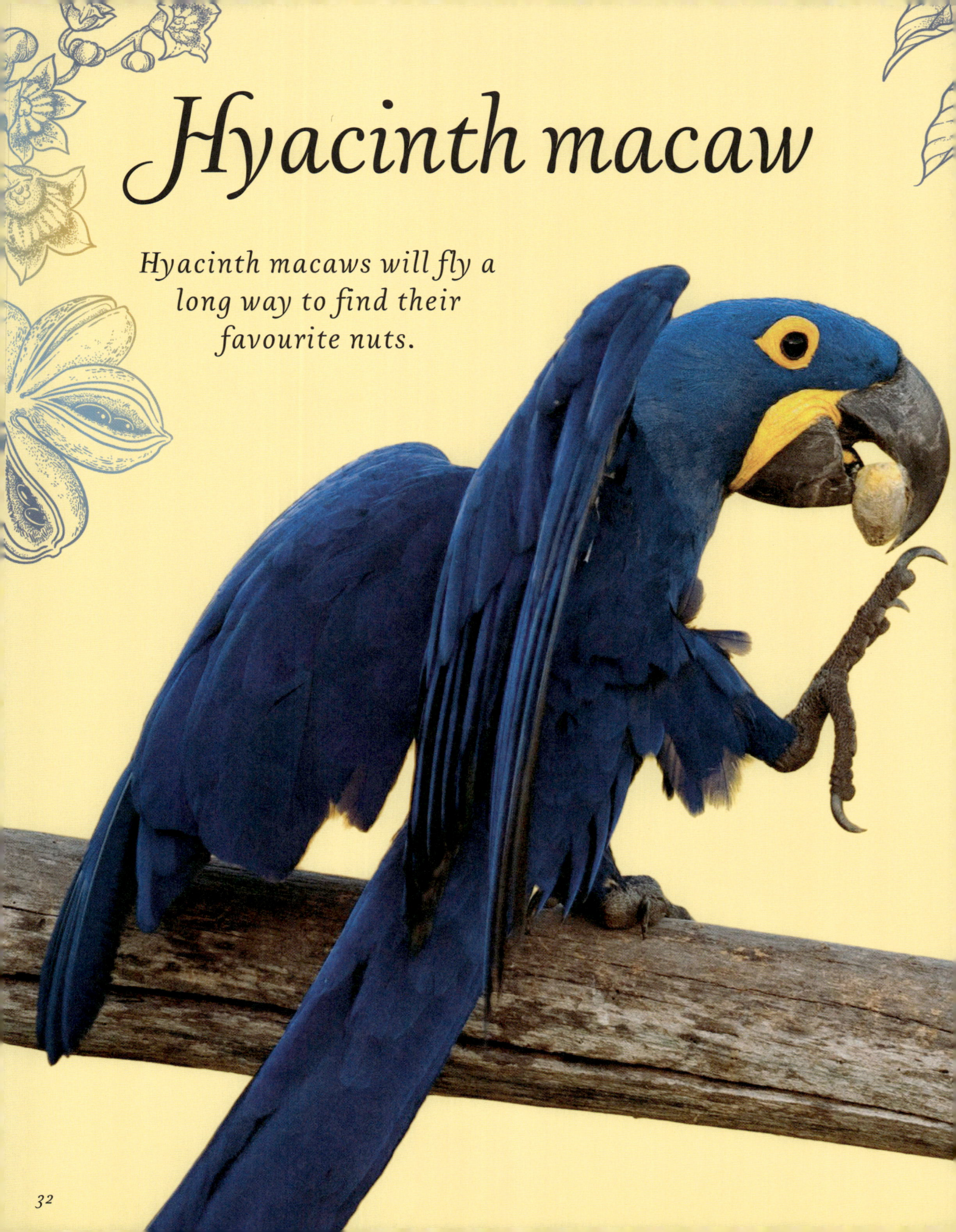

These spectacular birds are the biggest parrots on the planet. They have stunning blue feathers, with bright yellow skin around their eyes and beak. Even so, you will normally hear the macaws before you see them, because of how noisy they are. Their "kraaak kraaak!" calls echo far and wide over the woodlands and wetlands where they live.

Hyacinth macaws have a monster beak — if they really wanted to, they could crush your finger with it. But they are gentle birds, and their beak is actually adapted to crack tough nuts open to help them reach the tasty seeds inside. Like many parrots, the macaws form strong bonds with their partners. Each pair of macaws stays together for life.

Hyacinth macaw, South America. Paired macaws go everywhere together, and they love to preen each other's feathers, but not necessarily share food!

Great frigatebird

Male great frigatebirds have a brilliant party trick. To attract females in the breeding season, they blow into a pouch of skin on their throat. It fills with air, like a big, red love heart. They also drum on it with their beak.

Frigatebirds live at sea and nest in large groups on islands. Fish are their main food. Often the birds will fly behind dolphins, then snatch the fish that leap into the air to escape them. The frigatebirds are such superb fliers, they can travel far from land for weeks and weeks. They even sleep in mid-air. They never rest on the sea water, because their feathers aren't waterproof. One frigatebird flew over the ocean without landing for 60 days!

Female

Male

Great frigatebirds chase other seabirds to plunder their fish, a bit like feathery pirates.

Great frigatebird, tropical oceans. Male frigatebirds have an inflatable pouch under their beak, which females lack.

Lesser flamingo, Africa and Asia. Parent flamingos produce a milky liquid in their throat to feed their chicks. It's bright pink!

Lesser flamingo

Lakes in Africa sometimes look like shimmering pink seas. That's because the water is full of lesser flamingos! These pink water birds feed and breed in large groups. At certain times of the year, you can see up to a million of them on the same lake.

The flamingos have a bent beak, which they hold upside down and swoosh through the water. Little bits of food get trapped in the beak, especially tiny plants called algae — these give the birds their fabulous pink colour. Flamingos often feed at night, then spend the day resting. You will see them standing on one leg, with the other tucked in. It looks awkward, but saves them energy.

Flamingos have webbed feet and can swim like ducks.

Agami heron
Herons are experts at catching fish, and the agami heron is no exception. These birds wait patiently for a fish to swim near them and then dart their long, sharp beak down to catch it.

Evening grosbeak
The evening grosbeak has a wide, triangular beak to crack open tough nuts and seeds. Small grooves in the beak hold the nuts steady as the beak applies enough pressure to crack them.

Steller's sea eagle has the largest beak of any bird of prey.

Beaks

A bird uses its beak, or bill, to collect food, preen its feathers, and carry things, such as nest material. Having a beak is a key characteristic of birds, although turtles and tortoises also have one. In both birds and reptiles, the beak is made from plates of hard bone with a covering of keratin. Every bird's beak is adapted for its particular lifestyle, and as you'll see here, this means they come in a huge variety of shapes, sizes, and colours.

Hooded crow
Crows are found in many habitats and eat a wide variety of foods. They have a multi-purpose beak that can deal with small prey, the remains of dead animals, seeds, nuts, fruit, and much else besides.

Greylag goose
Geese feed on grass and seeds. The inside edge of their beak often has sharp ridges, like a saw, to help them shred mouthfuls of juicy grass.

Steller's sea eagle
All birds of prey have powerful beaks with a sharp, curved tip that can pierce and pull apart meat into smaller pieces to be swallowed. Steller's sea eagles use a pointed beak to tear up large fish, such as salmon.

Acorn woodpecker
You will often hear woodpeckers before you see them. To attract mates, they hit their strong beak on tree trunks to make a loud drumming sound. They also hammer on trees and drill into the bark to unearth insects to eat, or to make holes in which to build their nests.

Black skimmer
Skimmers have unusual beaks where the top half is much shorter than the lower half. As they fly over the ocean, they dip the longer lower half into the water to feel for fish, which are quickly snapped up.

Garden sunbird
The favourite food of sunbirds is nectar. They have a long, curved beak that can fit neatly inside the petals of a flower so they can reach this sugar-rich liquid. As they move between blooms, the sunbirds transfer pollen from flower to flower, pollinating them.

The curlew's beak acts like a long pair of tweezers to pick prey out of the mud.

Eurasian curlew
Carefully stepping over the wet ground of the seashore, the curlew repeatedly pokes its beak into the mud. It is feeling for small animals, such as worms and crabs. If it touches one, it quickly gobbles it up.

Pink robin
Most small perching birds eat insects and spiders, and for this diet a small, pointed beak is perfect. The pink robin of Australia has such a beak, which it uses to snatch prey off leaves.

Superb lyrebird

Female   Male

Lyrebirds can imitate almost any sound they hear.

There is something rustling in the undergrowth of the forest. It's a male superb lyrebird. Suddenly, he jumps onto a mound of earth he has made. This is a stage for his courtship display — and a female lyrebird has come to watch. The male lyrebird opens his tail to show off the lacy white feathers, and a gorgeous pair of curved feathers. Now he starts to sing, a mix of whistles, whoops, cackles, and squawks. His song sounds like it was created on a computer! But actually he copies most of the noises from other birds and animals in the surrounding forest. If the male's display wows the female, the two lyrebirds may become a pair and raise a family.

Superb lyrebird, Australia. Superb lyrebirds are so named due to their 16 tail feathers, which resemble the shape of a lyre — a musical instrument — when raised.

Great cormorant

Great cormorants LOVE fish. In fact, they eat nothing else. But sometimes the fish they catch are so big, you wonder how they will swallow their prey. Luckily, the cormorants have a stretchy throat to help them gulp their meal down whole. These birds swim beautifully, and barely make a splash as they dive from the water's surface to go fishing.

You can see great cormorants at lakes, rivers, marshes, and coasts. As long as there are plenty of fish, they are happy. The cormorants nest in groups, usually in trees beside the water — their frequent pooing soon turns the trees white! But they will also build their messy, smelly nests on the ground.

Cormorants swim low in the water — often all you can see is their head and neck.

Great cormorant, Africa, Asia, Europe, North America, and Oceania. Great cormorants have no problems swallowing a fish bigger than their own head.

Great northern diver

Divers can stay below the surface for three minutes.

A ghostly cry drifts across the water... It sounds like someone wailing. But there is no ghost — the strange cry is a great northern diver calling to its breeding partner. These elegant birds live on lakes, where they dive for fish that they catch in their spear-like beak. Their webbed feet are at the back of their body, near their tail, perfect for pushing them underwater at great speed as they chase their prey. Unfortunately, this means the birds are incredibly clumsy on land! They can hardly walk, so have to hop and shuffle.

Divers nest on islands in the middle of lakes. After breeding, they head to the coast for the winter. You can often see them swimming in the surf, not far from the shore.

Great northern diver, Europe and North America. The fluffy diver chicks are able to swim when just a few hours old.

Seriemas have such long legs, they seem to be on stilts.

Red-legged seriema, South America. With their long legs, these birds are excellent runners, and have a top speed of up to 40 kph (25 mph).

Red-legged seriema

Millions of years ago, South America was home to terror birds. These fearsome predators were 3 m (10 ft) tall, with beaks like eagles and claws like tigers. They were the bird version of Tyrannosaurus rex! Terror birds are extinct now, but if you visit the plains of Brazil and Argentina today, you can see their descendants: the seriemas. These birds are much smaller, of course, yet they are still fierce predators.

Seriemas mainly eat lizards, snakes, and other birds, and they chase these on foot. To kill their prey, they pick it up and bash it against the ground. Farmers often keep seriemas as "guard dogs" to protect their chickens — when the seriemas see a hawk or a fox, they shriek loudly.

Rhinoceros hornbill, Southeast Asia. Hornbills feast on forest fruit, especially juicy wild figs.

Rhinoceros hornbill

You can see how rhinoceros hornbills get their name! These rainforest birds have a huge horn, like that of a rhino, on top of their beak. It's bright orange-and-yellow, and hollow inside, which makes their calls louder. When the hornbills call to each other, their honks and barks can be heard a long way away.

Hornbills have really strange nests. The female hornbill climbs into a hole in a tree trunk and seals the entrance with mud and her own poo. Then, she lays her eggs inside the hole and raises the chicks in the dark. The male passes them food through a narrow opening. Eventually, the female and the young hornbills break out of the nest and emerge into the rainforest.

In flight, hornbills make a loud "whoosh whoosh" sound. Their wings are enormous!

Ring-necked pheasant, Asia, Europe, and North America. The male pheasant has colourful plumage, while the female is a dull brown.

Male   Female

Pheasants

Pheasant chicks can feed themselves from the moment they hatch.

Temminck's tragopan, Asia. To attract females, the male tragopan inflates the blue-and-red skin on his neck.

Himalayan monal, Asia. The male Himalayan monal shows off his finery in a mating dance.

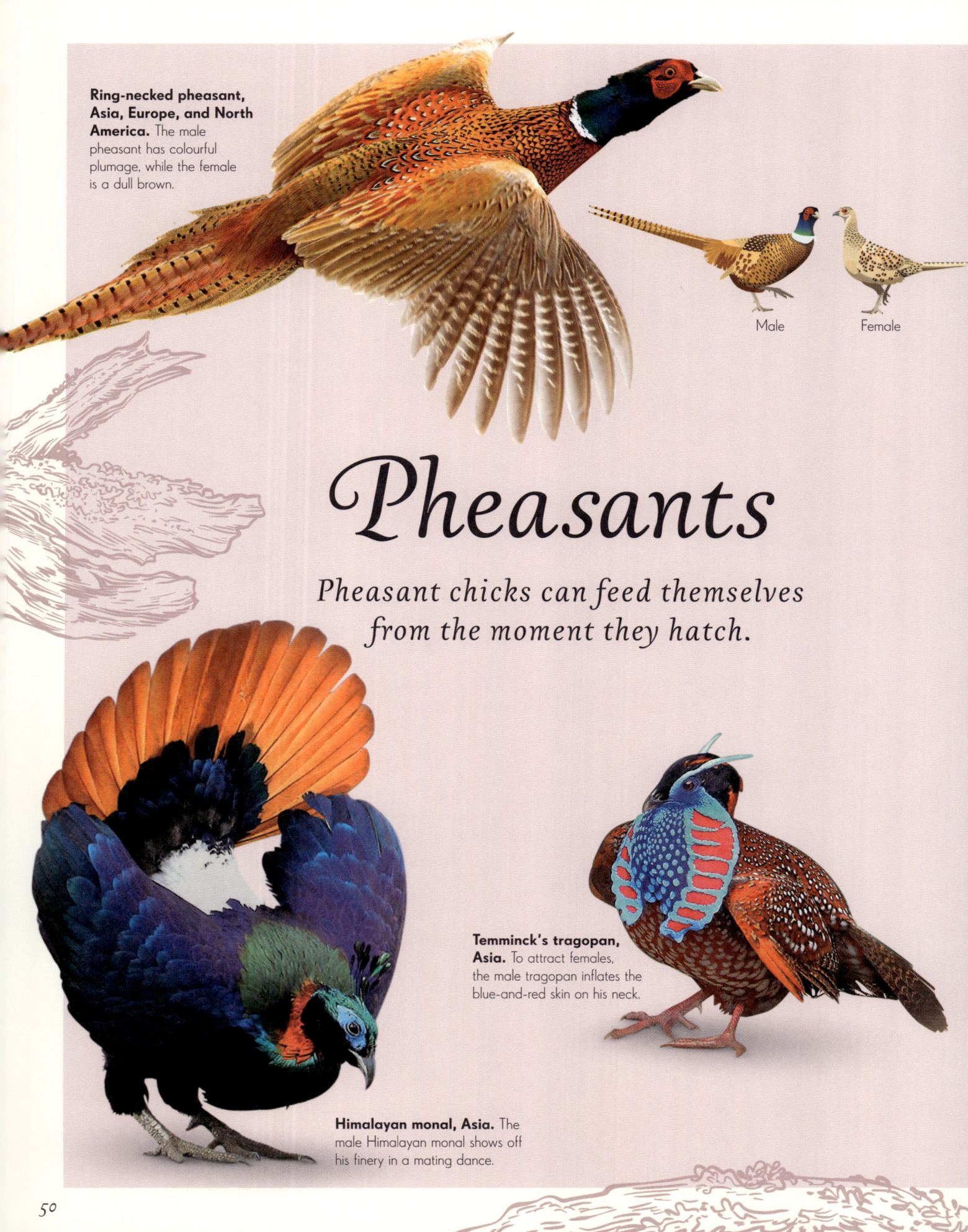

Lady Amherst's pheasant, Asia. This pheasant looks extraordinary, but it is incredibly shy.

Pheasants know how to put on a show! When the males stage their displays, they strut, bow down, or jump into the air. They often puff out their chest, or they might fan out their wings, or tail feathers. These dazzling performances are all about competing to attract mates. The females are less colourful, because they need to stay hidden on their nests.

Forests, mountains, and grasslands are where you will usually find pheasants. Though they can fly, they prefer to walk or run. They are related to chickens and turkeys, and like them, have strong legs. Male pheasants also have a spike, called a spur, on the back of their legs, which they use to fight.

Great argus, Southeast Asia. In his courtship display, the male great argus spreads his wings, like huge fans.

Anhinga

Anhinga, North and South America. The anhinga tosses its prey in the air, then catches and swallows it head-first.

An anhinga uses its sharp beak to stab and spear fish.

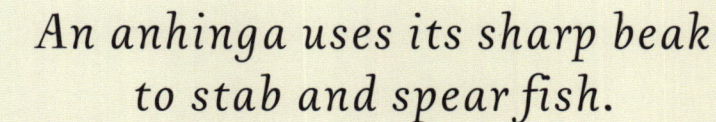

Swamps and marshes are the home of the anhinga. This curious water bird has a streamlined body and webbed feet like a duck, and it's an expert fisher. It often swims with its body underwater, and just its thin head and long neck poking above the surface. As most of the anhinga is hidden, it looks like a snake! No wonder the other name for this species is "snake bird".

Unlike other birds, the anhinga doesn't produce any special oil to waterproof its feathers. So, it emerges from the water completely soaked. Every time it goes fishing, it has to stand on the bank afterwards and hold its wet wings open to dry them out in the sun, like someone hanging out their washing.

Roseate spoonbill

You would probably find it difficult to eat with a spoon held in your mouth! However, for spoonbills, a spoon-shaped beak works brilliantly. The birds dip their unusual beak into water, open it slightly, and swing it from side to side. Whenever the beak touches prey, it snaps shut. The spoonbills mostly eat tiny shrimps and other small water creatures, though they can also catch crabs and fish.

There are six species of spoonbill, all of which live in wetlands or on the coast. Roseate spoonbills are the only pink species, and in the breeding season, their bald heads turn yellow and green. When in groups, spoonbills clack the two halves of their beak together to greet each other.

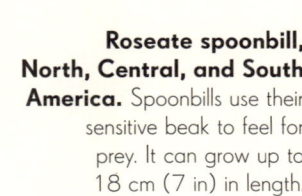

Roseate spoonbill, North, Central, and South America. Spoonbills use their sensitive beak to feel for prey. It can grow up to 18 cm (7 in) in length.

Young spoonbills have a straight beak. The full spoon shape develops later.

Male Female

Boobies pant like dogs to cool down in the hot sun.

Blue-footed booby, North and South America. A booby hatches with brown feet, but they turn blue as it grows up.

Blue-footed booby

The male blue-footed booby has the fanciest feet around. To attract a female, he dances in slow motion, lifting them up one at a time to show her. The brighter his feet, the stronger and healthier he is. These seabirds breed on remote islands. The female quacks to her mate, and he whistles back. You can see where they have been nesting, because they leave white circles of poo on the rocky ground.

Boobies fly out to sea to dive for fish. Wings folded, they speed into the water like missiles. On land, they have no fear of people. They shuffle about in a way that seems quite clown-like, which is how they got the name "booby". It comes from the Spanish word "bobo", meaning stupid or silly.

Snow goose, North America. There can be many thousands of snow geese in a single flock.

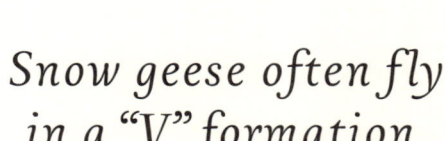

Snow geese often fly in a "V" formation.

Snow goose

Most snow geese have pure white bodies. When a flock of these birds flies overhead, it's like being in a blizzard! In North America, groups of snow geese are a sign of the changing seasons. Every spring, people look up to see the geese fly north to the Arctic to breed. There are plenty of plants to keep them fed there and many pairs of geese are able to raise four or five young.

In August, the snow geese fly back south for the winter. They have a long journey ahead of them, so stop at lakes and fields along the way to rest and feed. People look forward to seeing the snow geese again. When they arrive, it means winter is on its way.

A great bittern's "boom" can be heard 5 km (3 miles) away.

Great bittern, Africa, Asia, and Europe. Thanks to its long toes, this bird can climb reeds, and walk on mud.

Great bittern

Great bitterns would be amazing at hide and seek. They are shy birds that live in marshes and bogs, where they creep through the thick forest of reed plants that grow in the water. The streaky patterns on their feathers look just like reeds. If in danger, they stand still and point their beak up to the sky, which makes them "disappear" from predators.

Bitterns belong to the heron family. Like other herons, they have long legs to wade through water and a spear-like beak for fishing. In spring, the male great bittern makes a booming sound to establish his territory. You can make a similar sound by blowing over the top of an empty glass bottle!

When male junglefowl call loudly, it is known as "crowing".

Male

Female

Red junglefowl, Asia.
Male junglefowl have bony spikes called spurs on their legs, and use them to fight.

Red junglefowl

Does this bird look familiar to you? It's a red junglefowl, the wild bird that the domestic chicken was first bred from thousands of years ago. Red junglefowl come from forests in Southeast Asia, where they still live in the wild. The female junglefowl are mainly brown, to provide camouflage when on their nests. However, the males have colourful feathers, as well as wobbly red skin on their heads.

Domestic chickens are the world's most common bird. There are four times as many chickens as humans! Tests have shown that they are intelligent. They dream, have great memories, and can solve simple puzzles. So, when we raise these birds for their eggs and meat, we should do our best to treat them well.

Female brush turkeys lay up to 24 huge eggs.

Australian brush turkey

Birds need to keep their eggs warm until they hatch — usually, they sit on them. But Australian brush turkeys don't. Instead, they make an incubator for their eggs. They kick masses of leaves and earth into a mighty mound, which can weigh as much as several cars! Then the female brush turkey lays her eggs inside it. As the leaves in the mound rot, this creates heat, keeping the eggs warm.

The brush turkeys check their mound every day. If it's too cool inside, the birds pile more leaves on top. If it's too hot, they take leaves away. After around 50 days, the chicks hatch and dig their way out.

Australian brush turkey, Australia.
Brush turkeys are large, chicken-like birds with a bald red head.

Scarlet ibis

Ibises are water birds with long, curving beaks. Most ibises are white, brown, or black, but there is one species that is different from all the rest. Scarlet ibises are so red that their feathers appear painted. Their astonishing red colour comes from the crabs, shrimps, and other crustaceans that they eat — and the more of this food they consume, the redder they become. These ibises also eat a lot of beetles and frogs.

Scarlet ibises live in tropical wetlands and swamps, where groups of them nest in trees. From a distance, a tree full of these colourful birds can look like it is covered in exotic fruit! Young scarlet ibises are brown, then slowly turn red as they grow their adult feathers.

Scarlet ibis, South America.
Apart from black wingtips, scarlet ibises have entirely red feathers.

Young scarlet ibises can swim before they can fly.

Owls are whisper-quiet when they fly, thanks to soft wing feathers.

Great grey owl

Great grey owl, Asia, Europe, and North America The great grey owl's huge, round face looks like a satellite dish. It captures sounds and directs them towards its ears.

With their incredible night vision and hearing, owls are able to hunt at night. They can locate prey when it's much too dark for us to see anything. Some owls, like the great grey owl, are also active during the daytime. This is one of the largest — and fluffiest — owls in the world! It lives in cold northern forests and its thick feathers keep it warm in winter. Voles, mouse-like creatures, are its usual prey. When the owl hears a vole, it launches from its perch and swoops silently to grab the animal in its sharp claws. Even voles living in tunnels under deep winter snow are not safe! The owl can still hear them move and dives into the snow feet-first to seize them.

Sail feather
One unusually shaped feather is the sail feather of a male mandarin duck. Each male has a pair of these triangular, orange feathers, that poke up above his back, like tiny sails. They are part of a colourful display, used to impress females.

Contour feather
Contour feathers cover most of a bird's body. They are fluffy at the bottom, for warmth, but smooth at the top, to help create a streamlined shape. Usually, only the top part is colourful as that is the only part that can be seen.

Semiplume feathers don't have a continuous smooth outline.

Semiplume feather
Semiplume feathers are a bit like contour feathers — they are fluffy at the bottom and smoother at the top — but they have gaps between the strands of the feather. This type of feather often isn't visible, but helps to keep a bird warm.

Wing feather
Attached to the bones of a bird's arm, flight feathers form the wing. Wing feathers need to be strong to help flying birds push against the air to stay aloft. Many flight feathers are brightly coloured for use in displays.

Curled feather
Some bird feathers are a completely different shape to regular feathers. The male king bird of paradise has very long, wire-like tail feathers that are topped by a vibrant green swirl. He raises these above his head when displaying to females.

The Victoria crowned pigeon has a spray of filoplume feathers on the back of its head.

Filoplume
Some feathers are thin and wispy with a little fan at the end. These feathers are known as filoplumes. They are used in displays to mates or for sensing objects near a bird.

Down feather
Down feathers are light and fluffy, and no use for flying. Instead, they act like a cosy blanket underneath a bird's other feathers to keep it warm. When chicks hatch, they are covered in down feathers.

Types of colour
Birds come in every colour that humans can see, and some that we can't! However, some feather colours are a trick. Although they seem bright, the colours aren't produced by pigments, but by the structure of the feathers themselves.

Feathers

Feathers might be the most important feature of birds. All birds have feathers, and they come in many different shapes and a rainbow of colours. In birds that fly, feathers are the secret to keeping them in the air, but feathers do lots of other wonderful things too. They keep birds warm and waterproof, help them to communicate, streamline their body, sense touch, and can even produce sound. They are also the perfect fluffy lining for some birds' nests.

Iridescent feathers have a shiny satin sheen.

Camouflaged feather
While many birds are brightly coloured, others seem rather dull. However, a complex pattern of brown feathers can work as the perfect camouflage from predators. Female birds are often camouflaged to help hide them on their nests.

Tail feather
Male peafowl, called peacocks, have very beautiful tail feathers. Each long feather ends with a bright target of blues and greens that look a bit like an eye. Peacocks display these tail feathers to females to try and impress them.

Common raven

Ravens are the biggest crows in the world, with long wings and a powerful beak. They have a gruff voice, and their throaty "kronk kronk" cry is nothing like the cawing of smaller crows. The male and female form a strong relationship, and love to fly side by side. As a team, they soar, glide, and dive through the air together.

In captivity, ravens have shown that they can solve puzzles, and count up to seven. Ravens are playful birds, too, which is another sign of intelligence. They've been seen sliding down snowy slopes, seemingly just for fun. Then they return to the top and do it again. Who would have thought it — these birds enjoy sledging!

Ravens are fond of rolling over in mid-air, to fly upside-down.

Common raven, Africa, Asia, Europe, and North America. Ravens give gifts to their mate, often of food or nesting materials.

European herring gull

European herring gull, Europe. Herring gulls have many different cries, including wails and squawks.

People often call these big birds "seagulls". Herring gulls do live by the sea, but they live inland too. In towns and cities they nest on roofs, while at the coast they breed on cliffs. Herring gulls will eat just about anything. They catch small animals and search through rubbish. These intelligent birds even watch us in the street, ready to snatch any food we drop. We may find this behaviour annoying, but maybe we should also admire herring gulls for how smart they are!

Adult herring gulls have a red spot on their beak, which their chicks peck at if they are hungry. The chicks keep pecking, until the parent gulls bring up a sloppy grey meal of half-digested food. Mmm!

Gulls make lots of noise to defend their nest, raising around three chicks every year.

Like a cow, the hoatzin digests its leafy food with the help of bacteria.

Hoatzin

In the swamps of South America, you may meet a bird so odd, it is like no other bird alive today. The hoatzin can't walk and can barely fly. Weirdly for a bird, it eats leaves. Plenty of other animals munch leaves, but hardly any birds do. Because of this unusual diet, the hoatzin has sludgy poo that smells like cow dung. Local people call it the "stink bird"!

Hoatzin chicks have a claw on each wing. If they are in danger, they drop from their nest into the swamp water below. When the coast is clear, they use their wing-claws to climb back up again. The young hoatzins lose the claws as they grow up.

Hoatzin, South America.
A clumsy bird, the hoatzin uses its wings to balance.

Wild avocados are the favourite food of quetzals.

Resplendent quetzal, Central America.
A quetzal's silky green tail feathers are called a "train".

Resplendent quetzal

In misty mountain forests, there live glittering, green-and-red birds called resplendent quetzals. They are the size of pigeons, but have a stupendous tail, which in males can be over twice as long as their body! Living with a tail like this isn't easy. To take off, the quetzals have to step off their perch backwards. When they sit on their nest, they wrap their tail around them like a scarf.

Long ago, quetzals were worshipped by the Mayan and Aztec peoples. To them, these birds were the gods of the air. Their rulers wore headdresses made from quetzal tail feathers, which were taken from birds that were then set free, so their feathers could regrow. Today, quetzals are protected, and birdwatchers travel from all over the world to see them.

Kākāpō

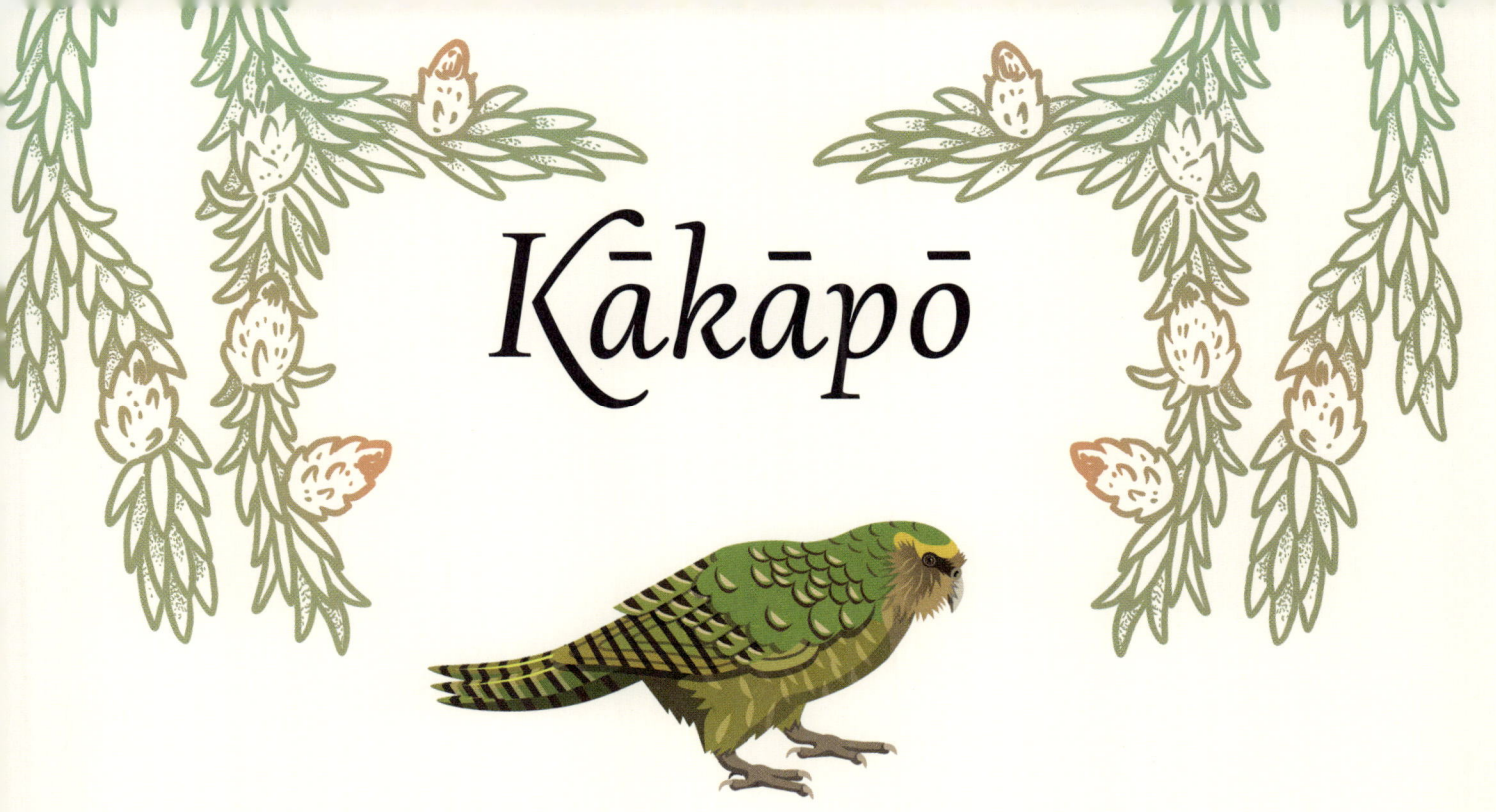

The kākāpō is a parrot, but a very odd one! It's the only parrot that can't fly, and is so big and heavy, it has to waddle everywhere. Unlike most other parrots, it is active at night, and its face is huge and round, like an owl. It has the softest, greenest feathers you can possibly imagine, and smells a bit like honey, or old wooden furniture. Male kākāpōs compete with each other to make loud booming noises. Their booms echo through the night and attract females, which will choose only the best boomers as their mates.

Sadly, kākāpōs are some of the world's rarest birds. Only around 250 survive today, mainly on islands near New Zealand. There is a huge conservation effort to save this extraordinary species.

The kākāpō can live for 90 years, and may have the longest lifespan of any living bird.

Kākāpō, New Zealand.
The green feathers of the kākāpō look just like moss, and are great camouflage in the forest.

Helmeted guineafowl are sacred birds in parts of Africa.

Helmeted guineafowl

Do you see that weird horn on the guineafowl's head? It looks just like a helmet! Long ago, some dinosaurs had a feature very similar to this. The structure is actually made of bone, covered with the same tough stuff your fingernails are made of. It's thought that the guineafowl may use their "helmets" to show off and display to each other.

Helmeted guineafowl live in groups most of the time. These birds like to dash around, as if always in a hurry. When something gives them a fright, they kick up a huge fuss and sound the alarm at the top of their voice. Their noisy calls sound like a bunch of people cackling.

Helmeted guineafowl, Africa.
These birds have a bare head. Loose bits of skin, called wattles, hang from their face.

Toucans often have play fights with their beaks.

Toco toucan, South America.
You can see toco toucans in all kinds of open country where there are trees.

Toco toucan

Everyone loves the toco toucan! This bird is famous all over the world, thanks to its super-sized beak that looks like a huge banana. The beak is hollow and full of air, so isn't as heavy as it appears. Otherwise, the toucan would topple off its perch. You might think a beak like this would be clumsy, but in fact, it's great for reaching fruit at the ends of branches. It also comes in handy for catching insects, lizards, and other prey. Toucans can even heat up and cool down their bumper beak, controlling their body temperature like a radiator.

Toucans travel in small groups. The birds make peculiar croaks and quacks to stay in touch with each other. They sound like frogs or ducks, only much louder!

Greater roadrunner

Roadrunners are fast on their feet. They have strong legs and can sprint at 30 kph (19 mph)! But why do they move so fast? It's because they race after their prey, mainly lizards, snakes, and mice. Snakes are not easy prey to overpower, but roadrunners smash them against the ground to stun them. Sometimes, the male and female roadrunner work as a team. One bird distracts the snake, while the other attacks it.

Roadrunners make their home in deserts and other dry places. They never drink — they obtain all the water they need from their food. Their feet have two toes pointing forwards and two backwards. So, their footprints look like lots of little "X" marks in the sandy earth!

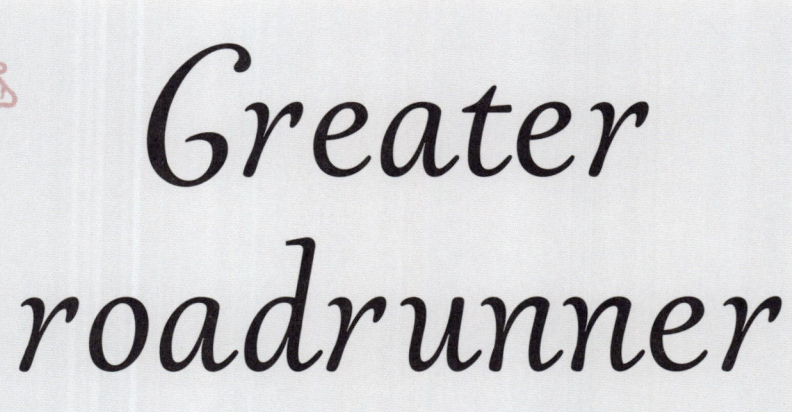

Greater roadrunner, North America. Roadrunners are built to run. They rarely fly, except when in danger.

At top speed, roadrunners use their long tail to steer.

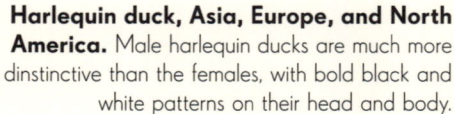

Harlequin duck, Asia, Europe, and North America. Male harlequin ducks are much more dinstinctive than the females, with bold black and white patterns on their head and body.

Harlequin duck

Ducks are familiar birds on ponds and lakes, even in towns and cities. One place you perhaps don't expect to find them is on wild rivers where the rushing water is white with foam. Yet, this is where harlequin ducks live. These ultra-tough ducks breed on ice-cold rivers in northern parts of the planet. Amazingly, they can dive to the riverbed to catch insects, water snails, and other creatures. The ducks are able to swim in powerful currents that would sweep a human away.

After breeding, harlequin ducks fly to the coast for the winter. But they don't go to sheltered bays with calm waters. No, they prefer to be on rocky shores, where the sea is rough and stormy.

Female Male

Harlequin ducks squeak like mice! One of their nicknames is "squealer".

Great crested grebe

Imagine being able to dance on water! Well, great crested grebes can. They live on lakes, where the male and female dance together to form their bond. As the pair of birds swim to face each other, they waggle and flick their head to show off a frill of feathers. People used to hunt the grebes for these feathers, which were worn on hats. The hunting stopped after people campaigned against it.

Grebes have sleek bodies with their feet at the back. This makes them strong swimmers, but means they can't walk on land. They dive to catch fish and they also eat feathers! The feathers stop fish bones from getting stuck in their stomach.

Grebes let their young chicks ride on their backs.

Great crested grebe, Africa, Asia, Europe, and Oceania. In one dance, the grebes rise up with water weeds in their beaks.

Cuckoo-roller

Cuckoo-roller, Madagascar. The female cuckoo-roller is brown with spots. The male is grey, with shiny green, blue, and bronze wings.

Cuckoo-rollers have a loud "whee-oo" cry, like a siren going off.

Sometimes, birds are so odd, we are not sure what sort of birds they are. Cuckoo-rollers are like this. They don't seem to have any close relatives among the world's other birds. For a start, they have a very big head, but their legs and feet look too small. And their eyes are in an odd position in the centre of their head. These birds really are a puzzle!

Cuckoo-rollers live only in forests on the island of Madagascar. Like many of the animals on this special island, they're not found anywhere else on Earth. The male and female cuckoo-roller stick close to one another in the treetops. Because they're always together, local people on the island often call them the "birds of love".

Montezuma oropendola, Mexico and Central America. Oropendolas weave sock-like nests from plant stems and leaves.

Montezuma oropendola

Montezuma oropendolas are tropical birds the size of crows. They have a patch of blue skin on their face, bright yellow feathers in their tail, and a most unusual beak, which is half black and half orange. Their nests look like socks hanging from branches. Oropendolas breed in groups in the same tree, usually near wasp nests. Why do they like wasps so much? Well, the stinging insects may keep predators away — and that means these clever birds can rear their young in peace!

Montezuma oropendolas have an incredible song, which rattles, bubbles, and gurgles. The species is named after emperors called Moctezuma, who were powerful rulers of the Aztec people.

Male oropendolas hang upside-down to sing.

White-nest swiftlet
The nest of the white-nest swiftlet looks like it might be made from plastic, however, the only ingredient in this nest is the bird's saliva! The swiftlets construct their bowl-shaped nests high up on cave walls, so the eggs inside are safe from predators.

Cliff swallow
Swallows build their nests using beakfuls of mud. The cliff swallow shapes its nest like a pear with a wide base and a narrow entrance at the top. Its nests are usually stuck to cliffs, but you will also find them under bridges and on walls.

Each blob of the nest walls is a beakful of dried mud.

Golden eagle
Eagles build humongous nests of branches. The golden eagle builds its giant nest on cliffs or at the tops of trees. A breeding pair will return to their nest and add to it each year, so it grows and grows.

Adélie penguin
The nests of Adélie penguins are not very comfortable. The birds gather together a small pile of rocks on which to lay their single egg. Sneaky rivals will try to steal the best rocks from other penguins when they aren't looking.

Common eider
The common eider's nest is super cosy. Each female plucks out fluffy down feathers from her chest to line her nest and keep it warm. Once the birds have finished nesting, people sometimes collect the soft eiderdown and use it in duvets!

Long-tailed broadbill
The dangling nest of the long-tailed broadbill looks a little messy, but it is a complex construction. A pair uses roots, stems, and dead leaves to create a sack-like nest, which hangs from a branch, or even a power line.

Nests

All birds need a safe place to keep their eggs, and many build a nest of twigs for this purpose, but not all. Some birds weave a nest of grass, others gather mud and build nests of clay, and a few even build nests from their own saliva! Eggs can't control their own temperature, so in order for them to stay warm enough to develop, most birds sit on top of the eggs in their nest and fluff out their feathers. This is called incubation.

Baya weaver
Weaver nests are some of the most complex. The male baya weaver carefully threads strands of grass together to create a chamber with an entrance tube that is difficult for hungry snakes to enter.

Anna's hummingbird
Hummingbirds build tiny nests that are very hard to see. Anna's hummingbird nests are not much bigger than a coin. They often disguise their nests with moss and lichen.

Song thrush
The song thrush builds a bowl-shaped nest from twigs inside a bush or other plant. It lines the inside with a layer of mud to make a smooth cup for its eggs to rest in.

Rufous hornero
It might surprise you to know this round mound of clay balanced on a tree branch is a nest. It belongs to the rufous hornero. These birds are also known as ovenbirds because their mud nests look a bit like traditional clay ovens.

The entrance hole to the nest is to one side.

Oilbird

Oilbirds spend the night searching the forest for fruit.

Oilbird, South America. Oilbirds look and behave like no other birds on Earth.

Deep in the rainforests of South America, there are caves where screaming sounds fill the air. These strange cries belong to oilbirds, which gather in these caves to nest. Oilbirds are active at night. Like owls, their eyes are incredibly sensitive, so they can see well in dim light. Like mice, they have long whiskers, with which they feel their way around. And like bats, they make clicking sounds to help them fly in total darkness. The oilbirds produce up to 250 clicks a second, and listen to the echoes to figure out what's around them as they fly through the cave and rainforest. This is called echolocation. It's like "seeing" with sound! Oilbirds are one of the only birds on the planet with this superpower.

Sunbittern

Sunbittern, Central and South America. The sunbittern has dazzling patterns on its wings that look like a pair of large eyes.

A sunbittern is hard to see – until it opens its wings.

One way to defend yourself from enemies is to pretend that you're dangerous. This is what the sunbittern does. It is a grey and brown bird and moves quietly, so most of the time, it blends into the background of its forest home. However, if something frightens the sunbittern, it suddenly spreads its wings to reveal bright patches of colour. Now the sunbittern seems to have a massive pair of staring eyes! The scary sight makes predators think twice about attacking the sunbittern and they leave the bird alone.

The sunbittern lives next to forest pools and streams. Although it is not a heron, it has a similar shape, as well as the same spear-like beak to catch prey.

Green heron, North and Central America. Green herons stay totally still, like a statue, as they wait for prey.

Some green herons use bread to attract fish.

Green heron

Catching fish is not as easy as it looks. However, green herons are skilful fishers. They stand or crouch at the water's edge and wait. The instant that fish swim into range, they seize the prey in their dagger-like beak. The herons can even hang upside-down off branches to catch fish below, and as they see so well in the dark, they also hunt at night. The green heron's cleverest trick, though, is to attract fish with bait, just like an angler. A heron picks up a suitable object to use as a tool, such as a leaf or a feather, and flicks it onto the water. When fish come up to eat the "food", the heron grabs its dinner.

Laughing kookaburra

In Australia, you may hear laughter when there's no one around. Don't be alarmed — it's only the laughing kookaburra. This bird's call is just like a human chuckle. When young kookaburras leave the nest, they remain on their parents' territory to help them raise their next brood of young. Often, the entire family burst into laughter together. The chorus of chuckles and shrieks is how they defend their territory from other kookaburras.

Laughing kookaburras are the largest members of a colourful bird family called the kingfishers. Some kingfishers live in wetlands and catch fish, but kookaburras prefer woods and farmland, as well as gardens and parks. Mostly they hunt lizards and snakes. Cheekily, they also steal and eat animals that the snakes have caught.

In populated areas, kookaburras pinch food off barbecues!

Laughing kookaburra, Australia. Kookaburras are bigger than city pigeons. Their chunky beak has a hook at the tip.

Northern potoo

Don't challenge a potoo to a staring match. It has a surprising trick — it can contract its pupils to reveal startling yellow irises! Like owls, it is a night hunter, and large eyes help it to spot tasty insects. Down in one they go, scooped into a massive beak that opens as wide as its head.

During the day, the potoo sits on a tree stump, stretches its head up, and freezes. The resting bird matches the bark perfectly and seems to vanish into thin air. To complete its disguise, the potoo keeps its eyes firmly shut, but it can still see what is going on, as there are small notches in its eyelids!

The potoo wails at night and is also known as the "ghost bird".

Northern potoo, the Caribbean and Central America. The potoo can open its eyes wide to startle predators.

Pileated woodpecker

In the forests of North America, you may see a crow-sized bird with a pointed beak and a strawberry-red crest. This is the pileated woodpecker. It clings to a tree and hammers on the trunk, its beak a blur. The woodpecker is after ants and other tasty insect prey hidden inside the wood. You can tell where it has been, because it leaves tree trunks covered in rectangular holes. In spring, it drums loudly on the trunks, hitting them with its beak around 15 times a second. Drumming is the woodpecker version of singing!

Though woodpeckers whack trees hard, they don't hurt themselves. Scientists think they may be protected by their powerful neck muscles and the shape of their beak, skull, and jaw.

Woodpeckers drum on hollow trees, as these make the most noise.

Pileated woodpecker, North America. Parent woodpeckers dig out holes in trees to raise their chicks – it's quite a squeeze inside.

Little spotted kiwi

Little spotted kiwi, New Zealand. Kiwis hunt at night for worms and other prey on the forest floor.

New Zealand is home to many curious creatures, including kiwis. These peculiar birds can't fly and have only stumpy wings. Instead, they scurry around forests at night, like rats or hedgehogs. Their feathers are loose and fluffy, so look more like hair. But the oddest thing about kiwis is their long beak. Unlike all other birds, kiwis have their nostrils at the far end of it. This means when kiwis push their beak into the soil, they're able to smell and feel their worm and insect prey underground!

Little spotted kiwis live mainly on a few predator-free islands. They are just as rare as the four other species of kiwi, but there is a particularly huge conservation effort in New Zealand to save these special birds.

Kiwis are constantly sniffing, to clear the dirt from their nostrils.

Little penguin

There are 18 kinds of penguin across the world, and these are by far the smallest. They stand around 30 cm (12 in) tall. As with all penguins, they can't fly, because their wings have evolved into flippers for speeding through water. Their feathers are small, and pack together to form a thick layer that keeps them warm and dry at sea.

We also call these penguins little blue, or fairy, penguins. Being small, they have many predators, so only come ashore after sunset, under the cover of darkness. They're not afraid of people, though. Some even nest near the Australian cities of Melbourne and Sydney. Once, a little penguin chick waddled onto an airport runway in New Zealand!

A little penguin can eat its own weight in fish every day.

Little penguin, Australia and New Zealand. These penguins bray like donkeys to greet their partners.

The peregrine falcon is the fastest animal on Earth.

Peregrine falcon, worldwide except Antarctica. A peregrine falcon can track its prey from 1.5 km (0.9 miles) away with its keen eyes.

Peregrine falcon

High in the sky, a peregrine falcon is on the hunt. Other birds need to watch out, because they are its main food. The peregrine seizes them in mid-air, often attacking from above. As soon as it locates a target, it folds its wings and dives. It can reach unbelievable speeds of up to 390 kph (240 mph) — faster than a racing car.

Truly, the peregrine falcon is a global bird. Not only does it live on six continents, it is also found in almost every type of habitat on land. It commonly lives in big cities, where it nests on bridges, cathedrals, and skyscrapers. This awesome bird has even worked out how to hunt at night, using the glow from buildings and streetlights!

Cape sugarbird

If you want to see sugarbirds, you first need to find some flowers. These birds live on nectar, so they visit dozens of flowers a day. Above all, they search for spectacular flowers called proteas. The tip of their long tongue is like a brush, which helps them sweep the sweet nectar into their mouth. The sugarbirds flit here and there to feed, and as they do so, they transfer pollen between different flowers, which pollinates them.

Cape sugarbirds live only in the Cape region of South Africa. The sunny hillsides in this area have a mind-boggling variety of wild flowers, including many proteas. When the protea flowers bloom, this is when food is easiest to find, and the sugarbirds raise a family.

Cape sugarbird, South Africa.
Sugarbirds love the nectar of protea flowers.

Sugarbirds use their long tail to balance on flowers.

Livingstone's turaco

Livingstone's turacos have dazzling colours and a spiky crest. So as you can imagine, many birdwatchers want to see these eye-catching birds. But it's not easy, because turacos are shy and live in Africa's thickest forests. Rather than fly, they like to hop around the treetops like squirrels, hidden among the leaves. The turacos hardly ever come to the ground, except to drink and bathe.

There's something very special about turaco feathers — their bright green and red feathers have copper in them. The copper is contained in rare pigments, which the birds get from fruit, their favourite food. In one popular tale, the turacos' wonderful colours wash out of their feathers in the rain. Happily, it's only a myth.

Turacos have unique feet, with a rotating toe to grip branches.

Livingstone's turaco, Africa.
This turaco is mainly green and blue, with a magnificent green crest tipped with white.

Bar-tailed godwit, Africa, Asia, Europe, North America, and Oceania. Godwits feed on worms and snails buried in mud on the coast.

Bar-tailed godwit

The bird you are looking at is a world champion. It may just have flown halfway around the Earth! Bar-tailed godwits nest in the Arctic at the top of the planet, but they spend the winter on muddy seashores far to the south. To travel between these distant places, the godwits make the longest non-stop flight of any bird. We know this because scientists attached tiny tags to the birds that send a signal back with their location.

Before the godwits set off, they eat a lot to put on fat, which their body will burn as fuel. They grow fresh wing feathers and their heart and flight muscles get bigger, ready for the tough journey ahead.

One godwit flew more than 12,000 km (7,450 miles) non-stop, over 11 days.

Snow petrel, Antarctica and Southern Ocean. The petrels can drink sea water, and filter out the salt through tubes on their beaks.

Snow petrel

You can see why this bird is called a snow petrel. With its pure white feathers, it seems to be made from snow itself. Snow petrels are much tougher than they look, though. They live in Antarctica and the icy Southern Ocean, the coldest part of the planet. Scientists have even seen snow petrels at the South Pole, where few other animals have ever been recorded.

Snow petrels flutter over the freezing sea to find food. Mostly they eat shrimp-like animals — as many whales do, too. The petrels land on icebergs to rest. If other birds come too close to their nest, they open their beak to spray the intruders with stinky, orange vomit!

These birds roll in snow to clean their feathers.

Namaqua dove, Africa and western Asia. These tiny doves are no bigger than sparrows, but have a very long tail.

Common bronzewing, Australia. You can see these plump pigeons in all kinds of habitats, except for thick forests.

Pigeons

What's the difference between a pigeon and a dove? It's a trick question — there isn't one! Pigeons and doves belong to the same big family of birds. They come in many shapes, sizes, and colours. Usually, they bob their heads while they walk, as if listening to music. Parents produce a rich liquid in their throat, similar to milk, to feed their chicks.

Some pigeons and doves have become rare. Pink pigeons, for example, were once down to just 10 individuals in the wild. But they were bred in captivity and now there are around 500 of these pretty pigeons. This heartwarming rescue story shows that we can often save rare birds — if we act in time.

Pink pigeon, Mauritius. Pink pigeons live only in forests on the island of Mauritius. They eat fruit, flowers, and seeds.

Pigeons are clever birds, and have even been taught to play table tennis!

Thick-billed green pigeon, southern Asia. A large beak allows these pigeons to feed on fruit, especially wild figs.

Nicobar pigeon, southern Asia. These dazzling birds have a "shawl" of long feathers around their neck and breast.

Rock dove, worldwide. These familiar birds are common in cities, but originally nested on sea cliffs.

Spinifex pigeon, Australia. Males and females of this species both have a tall, pointy crest.

Grey parrot

Grey parrots solve some puzzles faster than a five-year-old child.

A chorus of ear-splitting screeches can only mean one thing: parrots are in the area. Extremely noisy birds, they fly around forests and woodlands in flocks, on the lookout for fruit, nuts, flowers, and other plants to eat. Many parrots look dazzling, with rainbow-coloured feathers, but grey parrots are plainer than other species. They are, however, excellent mimics. In captivity, grey parrots often copy human speech, and some may learn as many as 150 words.

The scientist Irene Pepperberg studied a grey parrot, called Alex, for more than 30 years. She discovered that Alex could count and recognize different colours and shapes. He could even ask simple questions. As he loved being tickled, he would ask for more tickles!

Grey parrot, Africa.
Parrots have big brains for their size. They're among the cleverest of all birds.

Tufted puffin, Asia and North America. With their huge beak and fancy feathers, these birds are the parrots of the sea.

*A puffin can carry up to
20 fish lined up inside its beak.*

Tufted puffin

Tufted puffins are seabirds that look like they have dressed for a party! Their body is pure black, with a bright white "mask" on their face, and curly tufts of golden feathers on their head. But they're only this stylish when breeding. At the end of the summer, they swap their gorgeous feathers for dull black ones and the colourful bits of their beak drop off.

Puffins are tough birds that can survive wild ocean storms. Much like penguins, they dive far below the surface to catch fish, squid, and shrimp-like animals. Using their wings for power, they "fly" underwater. Adult puffins stay with the same partner for life. They nest on rocky coasts and islands, where together they rear a single fluffy chick each year, called a puffling.

Scissor-tailed flycatcher

Scissor-tailed flycatcher, North and Central America. These acrobatic birds can even do backflips in the air.

No other birds in the USA have tails like this.

Scissor-tailed flycatchers have a secret weapon: their tail! When they take off, their long tail feathers open like a pair of scissors. This helps them twist and turn to catch flying insects. Often they swoop to the ground to grab grasshoppers and beetles.

In summer, these charming birds are found in southern parts of the USA. They like to perch on roadside fences to get a good view of prey. Pairs of flycatchers defend a territory and will not put up with other flycatchers there. If they spot one, they chase it away. After breeding, the flycatchers fly south to Central America for the winter, where it is warmer. When spring comes, they make the return trip.

Eurasian coot

You can spot a coot just by its strange feet. All of its front toes appear swollen, but this is normal. The extra flaps of skin help these water birds paddle on lakes and ponds and squelch across soft mud. The flaps actually do much the same job as the webbed feet of ducks.

Coots can also be identified by the bright, white shield on their forehead. They may use these featherless areas to recognize each other, or in displays. During the breeding season, coots become very bad-tempered. Other coots who get too close to their floating nests are attacked and chased away.

Fighting coots kick out fiercely with their feet.

Eurasian coot, Africa, Asia, Europe, and Oceania. The wide flaps of skin on a coot's scaly feet stop it sinking in soft ground.

Arctic tern

Arctic terns glide over the ocean for hours at a time. Fantastic fliers, they can even hover over the waves. In common with many seabirds, they eat fish, dipping into the water with a splash to catch them.

No birds are better travelled than Arctic terns. They nest on seashores in northern parts of the world, as far north as the Arctic. Having raised as many as three chicks during the breeding season, the terns then fly to the Southern Ocean, as far south as the sea ice of Antarctica. For some terns, the return trip can be 95,000 km (59,000 miles). That is like flying around the planet two and a half times, and what's more, they do it again every year!

Arctic tern, oceans worldwide.
Arctic terns have long wings and a lightweight body.

Arctic terns enjoy two summers each year, so never see winter.

Common cuckoo

The common cuckoo is a cheat. It doesn't build a nest or raise its young, but tricks other birds into doing it instead! The female cuckoo sneakily lays her egg in the nests of others, such as the small reed warbler. Her eggs look exactly like those of the birds she picks on, so the real owners of the nests don't notice anything wrong. Once hatched, the baby cuckoo pushes out any other eggs and young to get the nest to itself. Now, the unwitting foster parents feed the cuckoo chick as if it were their own.

There are many different species of cuckoos around the world, and most of them do actually look after their own young.

Adult reed warbler

Cuckoo chick

Common cuckoo, Africa, Asia, and Europe. In spring, the male sings a "cuck—oo, cuck—oo" song.

Female common cuckoos lay 10 or more eggs, each in a different nest for other birds to look after.

Yellow-bellied prinia

The yellow-bellied prinia isn't a flashy bird – it is yellowish-green and smaller than a sparrow – but it lays gorgeous eggs! The colour of the eggs comes from a red pigment. This is one of only two pigments that, mixed together, produce all egg colours.

Wattled jacana

The wattled jacana is a wading bird that lays its eggs in a riverside nest of water plants. The eggs' squiggly patterns and yellow shell help to camouflage the eggs from predators.

The squiggles help to break up the outline of the egg.

Eggs

Eggs are some of the most beautiful objects in nature. Birds lay hard-shelled eggs in order to reproduce, and so do some reptiles – dinosaurs, the ancestors of birds, did as well. An egg is a safe home in which a chick can grow. The tough outer shell protects the developing baby, but also has tiny holes, called pores, to let in oxygen. The eggs of different birds can look wildly different. They can be almost any colour and many have intricate patterns on them.

Guillemot eggs are pointed at one end.

Guira cuckoo

Unlike some cuckoos, guira cuckoos look after their own eggs and chicks. When their eggs are laid, they look chalky white, but this layer flakes off to reveal a beautiful ocean blue colour beneath.

Common guillemot

These eggs look like someone has splattered brown paint over them. The wiggly markings are unique, and may help the parent birds spot their own egg in the large colonies where they nest.

Ruby-throated hummingbird

Hummingbirds lay some of the smallest eggs. The tiniest are around the size of a pea! Some hummingbird eggs will hatch in as little as two weeks.

Scissor-tailed flycatcher

Many birds lay eggs with spots or splotches. The eggs of scissor-tailed flycatchers look like spotty Dalmatian dogs. Like speckled or squiggly-patterned eggs, the spots probably help to hide the eggs in their nest.

American robin

It's hard to miss the bright blue-green eggs of American robins. Many species of bird lay blue eggs, which get their colour from a blue pigment called biliverdin. The brightness of the blue can vary between eggs.

Ostrich

Ostriches are the biggest birds in the world and they lay the biggest eggs. Each ostrich egg is around 15 cm (6 in) long and weighs as much as 24 chicken eggs!

Saker falcon

Like the eggs of many falcons, the eggs of saker falcons are heavily speckled to help camouflage them. These eggs are either laid in the used nests of other birds or on ledges on cliffs.

Elegant crested tinamou

The eggs of tinamous have a glossy surface, so almost don't look like birds could have made them. Different species of tinamou lay eggs of different colours, including pink, blue, brown, or in the case of the elegant crested tinamou, grass green.

White tern

White terns are small seabirds that breed on tropical islands. They lay a single egg, but don't make a nest — they simply balance it on the branch of a tree! Somehow, the parent terns sit on the egg without knocking it off the branch. The chick, when it hatches, has to hold on tight with its sharp claws, or it might fall off. Sometimes, the terns rest their egg on a rock, or even on the window ledge of a building. This seems dangerous, but keeps it safe from predators on the ground.

White terns swoop to catch fish at sea and can carry several at once to feed their young. When they fly above you, their pale wings are almost see-through.

These handsome white birds are also called fairy or angel terns.

White tern, Indian and Pacific Oceans. These terns lay their eggs in some very odd-looking places.

Ruff

Male Female

Female ruffs raise their young alone, with no help from the male birds.

Ruffs are great show-offs. Every spring, the male ruffs dance together in groups. It's like putting on a dance competition! But why do they go to so much effort? They do it to impress the female ruffs, known as reeves, which gather to watch. Only the best dancers will catch the eye of the females and get to form a pair.

Ruffs are a type of shorebird, with long legs for wading, and long beaks that they push into water, grass, and mud to find food. In spring, ready for their dance, the males grow frilly feathers on their head and neck. These look like the fancy collars, or ruffs, worn by wealthy people in Europe hundreds of years ago.

Ruff, Africa, Asia, and Europe.
The colour of a male's neck "ruff" can be white, black, golden, or rusty red.

Eurasian hoopoe

Hoopoes are named after their lovely song. In his beautifully soft voice, the male sings "poop poop poop" to mark out a territory and attract a mate. Both the male and female hoopoe have a magnificent spiky crest, made up of peach feathers with black-and-white tips. They raise and fan it to show they're excited. When they fly, their wings appear very floppy, so they flutter along like giant butterflies. They use their spectacular curved beak to catch beetles, small lizards, and other prey.

People have always been fascinated by hoopoes. The ancient Egyptians featured them in their paintings and picture writing system. They also appear in the holy texts of Muslim, Jewish, and Christian people.

Eurasian hoopoe, Africa, Asia, and Europe. Both hoopoe parents bring food to their chicks.

If hoopoe chicks are attacked, they aim a stream of poo at their enemy!

Andean cock-of-the-rock

As dawn breaks in the misty South American forest, you hear a mighty din. A group of bright orange birds have gathered in a tree and they're squealing like pigs. They seem agitated, fluttering about and jumping up and down on their perches. What's going on? These birds are Andean cock-of-the-rock, and all are male. Every morning, they turn up at the same tree to show off their finery. The females, which are brown, visit the tree to watch and choose the best performer as their mate.

Later, the female cock-of-the-rock head into the forest to bring up their family on their own. They build their nest on a rock face, which is why these extraordinary birds have "rock" in their name.

Male cock-of-the-rock look like they have half a tangerine on their head!

Andean cock-of-the-rock, South America. The male birds have a fan-like crest that hides their beak.

In autumn, a busy blue jay can bury up to 5,000 acorns.

Blue jay

Blue jays are a stunning blue colour... Or are they? These birds have a surprising secret — they are not actually blue, but brown! Just like the sky on a sunny day, their feathers only seem blue to us because of the way they scatter light particles.

Forests are the main habitat of blue jays, but they like city parks and gardens, too. Their loud screeches echo through the trees. They eat lots of things, especially seeds, and nuts such as acorns. During the autumn, they collect thousands of acorns and bury them to eat later in winter. Though the jays are smart birds and have a great memory, they always forget where some acorns are hidden. These missing acorns will one day grow into oak trees, which means the jays are helping to plant the forests of the future.

Blue jay, North America. The wings of blue jays have an exquisite black and blue pattern.

Bearded bellbird

Tropical rainforests can be noisy places. Everywhere you go, you will hear the calls and songs of many different birds. However, when male bellbirds perch on top of a tree and belt out their mating calls, you really know about it! These are the loudest birds on Earth. There are four species of bellbird, all of which sound different. The male bearded bellbird yells "chonk! chonk! chonk!" over and over again, which sounds like a bell chiming or a hammer hitting against metal.

Does the male bearded bellbird have a beakful of worms? Actually, it's a "beard" of stringy black feathers. The female doesn't have this freaky feature, and is green in colour. She's much quieter than the male, too — in fact, she hardly makes a sound.

Male bellbirds are as loud as police sirens.

Female

Male

Bearded bellbird, South America. A male bearded bellbird grows his "beard" of feathers when he is one or two years old.

Namaqua sandgrouse

Sandgrouse may fly 50 km (31 miles) to find water.

Life is tough in deserts and dry, dusty plains, but even here, there are birds. Namaqua sandgrouse are experts at living in these extreme conditions. Every morning, they gather in flocks and fly a long way to a pool or puddle to drink. Afterward, they spread out to search the sandy earth for seeds to eat. Their sand-coloured feathers match their surroundings, which hides them from birds of prey, their main enemies.

When they are little, sandgrouse chicks can't fly and they soon get thirsty. So the adult male sandgrouse do something amazing. After sitting in water to soak their belly feathers, they fly back to the nest and let their chicks sip their wet feathers to get a refreshing drink!

Namaqua sandgrouse, Africa. These unusual birds have a pigeon-like body, short legs, and a pointed tail.

Trogon parents often feed lizards to their chicks.

Cuban trogon

Trogons love to perch halfway up a tree and sit there very still, to watch the world go by. Until these birds move, they blend into the background of their forest home and seem to disappear. But when you do finally spot them, their gorgeous feathers are a feast for the eyes.

There are around 40 different kinds of trogon, all found in tropical forests. Cuban trogons can be seen only on the island of Cuba. Local people call them the "tocoroco", because of the curious sound they make. The trogons feed on a variety of fruit, flower buds, and insects. They nest in a hole in a tree trunk. Usually, they move into an old one already made by a woodpecker, which is much easier than having to dig it out themselves.

Cuban trogon, Cuba.
Male trogons open their beak, wings, and tail, in a threatening display, to defend their perch.

Killdeer

These little birds don't really kill deer! They are named after their "kill–deer" call, a common sound in much of North America. Killdeer belong to a group of birds known as shorebirds, and like their relatives, have long legs for wading in water — but they often live far from it. You can see them wherever the ground is grassy or stony, including beside roads, and even at airports, car parks, and building sites.

If a predator, such as a fox, sneaks up to a killdeer's nest, the bird plays a trick on it. The killdeer flutters along the ground and the predator follows, thinking the "injured" killdeer will be easy prey. Then suddenly, the crafty killdeer flies back to its nest, leaving the predator confused and empty-handed.

*Killdeer are unafraid of people
and can be very tame.*

Killdeer, North, Central, and South America. When faced with a predator, the killdeer twists its wings to pretend they are broken.

Flame bowerbird

Bowerbirds are some of the greatest natural architects.

Birds put a lot of effort into attracting a partner. Many sing, or grow fine feathers. Some dance, and others perform dramatic flying displays. Male bowerbirds impress the females with their building skills. They live in forests and build elaborate structures, called bowers, from sticks and leaves. The bowerbirds decorate their masterpieces with flowers, berries, shiny beetle wing cases, snail shells, and other eye-catching objects.

For his bower, the male flame bowerbird builds two neat walls of sticks. The opening between them is like an avenue, which he struts along, as if on a stage. The female flame bowerbird is fussy, however. She will tour the bowers of different males before picking the one she likes best.

Flame bowerbird, New Guinea.
The flame bowerbird's glamorous feathers look even better standing in his bower.

Long-tailed manakin

When not dancing, manakins are shy birds.

Long-tailed manakin, Central America. The colourful males have a pair of very long tail feathers. The females are green, with a short tail.

Few birds can dance as well as manakins. It is the males that dance, and their display is all about attracting females. Long-tailed manakin males dance together as a team. Their "dance floor" is a large, horizontal branch. On it, they perform several routines, including the "up-down" dance. One male hops up into the air and flutters back down to the branch, then the next male does it, and so on, until they have all joined in. They even sing at the same time, and leapfrog over each other's backs to swap places! The manakins spend at least five years practising these dance moves together.

European starling

A flock of starlings flying together is called a murmuration.

If you hear a phone ring up a tree, don't worry. It could be a starling! These smart birds copy all kinds of sounds, including ringtones, power tools, and other birds. Like composers, they mix the sounds they learn with their own chatter and whistles.

On winter evenings, starlings gather to roost. Huge flocks of them swirl around the sky before they settle down for the night. The starlings fly close together and keep changing direction, so it's tricky for birds of prey to catch any. As the flocks twist and turn, they create amazing patterns high above. Yet they never crash. Up to a million birds move as one, like a single giant creature.

European starling, Africa, Asia, and Europe. A cloud of starlings appears to swirl around the sky as the birds change direction together.

Common crow
Crows are aerobatic birds. They have wide wings that allow them to take off fast if they spot danger and change direction quickly. They also have slots at the tips of their wings to help them catch air currents.

Storks have finger-shaped feathers at the tips of their wings.

Indian roller
Rollers are named for the energetic flights that males make to impress females. During these spectacular displays, they seem to tumble and roll through the air. They wait until they spot prey, before swooping down to catch it.

White stork
Storks have huge, wide wings. They alternate between slowly flapping and soaring through the air. The feathers at the tips of their wings splay out like fingers to help catch warm air currents as they rise.

When they spread their wings, rollers show off their brilliant blue and turquoise feathers.

Wings

Every species of bird has wings, although some are flightless. A bird's wing shape is adapted to its particular lifestyle — from the long, sturdy wings of albatrosses that allow them to fly the equivalent of twice around the globe without landing, to the heavy wings of penguins that help them to dive as deep as 450 m (1,470 ft) in the ocean. By looking at the shape of a bird's wings, you can take a good guess as to how much of its life it spends in the air.

Black-browed albatross

Albatrosses have very long wings that are perfect for soaring over vast distances. They catch updrafts above ocean waves, so hardly ever need to flap their wings to stay airborne. Their wing shape means they can fly far using very little energy.

Pacific swift

The Pacific swift is shaped a bit like a jet plane. It zigzags through the air using its narrow, curved wings and swoops high to hunt insects. Like the common swift, this bird rarely lands, and even sleeps while flying.

Rufous-tailed hummingbird

Hummingbirds may be small but they are champion flyers. Their wings actually move in a figure of eight pattern, rather than flapping up and down, which allows them to push against the air continuously. They can hover, and even fly upside-down and backwards!

Barn owl

Owls rely on the element of surprise to catch their food. Their wings are designed to flap silently, so as not to alert prey below. The feathers at the back of their wings have soft fringes, to muffle the noise of air rushing over them.

Emperor penguin

Penguins are too heavy to fly, but their long, thin wings are as perfect for gliding through the water as they would be flying in the sky. Diving penguins flap their wings to swim, while a thick coat of waterproof feathers keeps them dry.

Northern cardinal

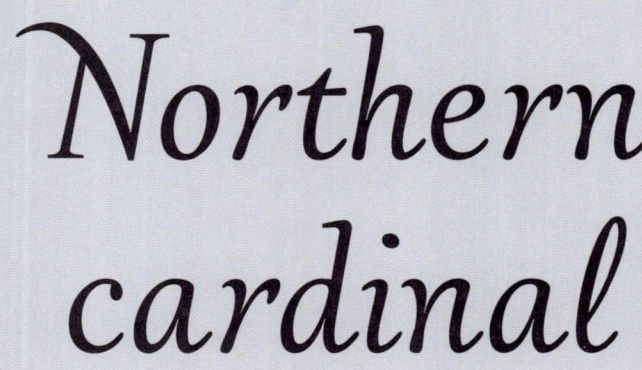

These gorgeous birds are as red as can be. Even their beaks are red. However, only the males look so bright. Females have pinkish-grey or brown plumage to hide them when they're nesting. But why are these birds called cardinals? They are named after officials in the Catholic Church, who wear red caps and capes.

These birds make a cheerful sight in gardens and parks, and love to visit bird feeders. Sunflower seeds are their favourite. In spring, the male perches beak to beak with his female partner to offer gifts of food. The pair often whistle a beautiful duet, which makes their relationship stronger still.

The cardinal is the state bird of seven US states and a mascot for many American sports teams.

Northern cardinal, North America. Most of a male cardinal's feathers have a red hue, except for those around the beak, which are black.

Noisy pitta

These birds are indeed noisy. For half the year, their loud whistles ring out in forests, day and night. At other times, they fall silent. The pittas hop around the forest floor to look for snails, beetles, and earthworms among the fallen leaves. Often, they have a favourite rock where they smash snails to break the shells and reveal the juicy flesh inside. The pittas may cover their nest with the poo of wallabies, a type of small kangaroo. This sounds gross, but it's a clever defence against snakes. The poo is so stinky, snakes can't smell where the nests are.

The name of these birds has nothing to do with pitta bread! It comes from an Indian language, meaning "small bird".

Noisy pitta, Oceania. Pittas have such a short tail, it looks like they haven't even got one.

Male and female noisy pittas are both as colourful as each other.

White-throated dipper

Some dippers build their nests behind waterfalls.

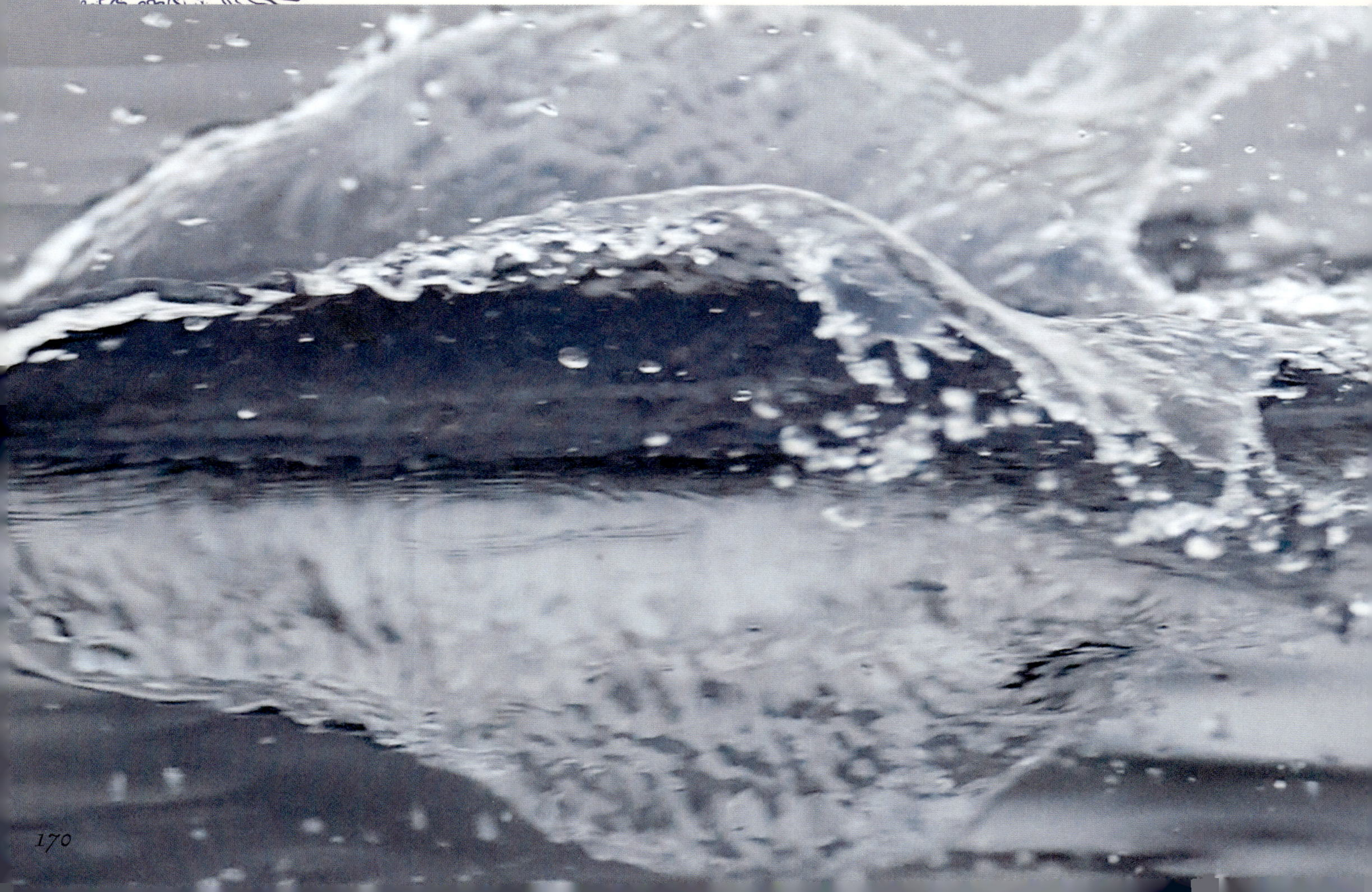

*I*magine being able to walk underwater. Well, these birds can! White-throated dippers live around mountain streams and rivers. As if by magic, they sit on a rock next to the rushing water, jump in, and disappear. Of course, they actually dive down to the stony riverbed. The dippers walk among the stones, searching for water insects, shrimp, and other prey. They can also swim underwater, with their wings beating to push them along.

Dippers have a round body about the size of an orange. They are the only birds this small that walk and swim underwater. Their name comes from a peculiar, unexplained habit of theirs. They like to sit on rocks and bob up and down as many as 50 times a minute.

White-throated dipper, Asia and Europe. With their wings whirring, dippers speed along just above the water.

Bohemian waxwing, Asia, Europe, and North America. Acrobatic feeders, waxwings use their wings and tail to balance.

Bohemian waxwing

A soft whistle tells you that waxwings are nearby. These beautiful birds have a peach-coloured body and spiky crest. In summer, they live in pine forests in the north of the world, catching flies to eat. But in winter, they switch to eating fruit. Some years, the trees in these forests don't produce many berries, so the hungry waxwings fly far to the south. They often head to urban areas, and you can even see them feasting on berry trees next to busy city streets. The traffic and noise doesn't seem to bother them. Some of their wing feathers develop a curious waxy red coating at the tip, which may be a sign that the birds are fully grown. The unusual waxy wings give these birds their name.

A waxwing can eat more than 800 berries a day.

Greater honeyguide

Bees' nests are full of food. However, they are hard to break into for a little bird like the greater honeyguide. So, it asks for help. The bird gives a special call, which attracts a honey badger, then guides it to a nest. The powerful badger rips the nest apart, and when it has finished eating, the honeyguide enjoys its share. It eats young bees and also the tough wax that bees use to build their nest. By working together, the bird and the badger both get a meal.

In parts of Africa, the honeyguide also teams up with people. It whistles to them and guides them to honey-filled nests. This is a rare example of a wild animal working with humans.

Honeyguides are the only birds able to eat beeswax, but still need to watch out for stings!

Greater honeyguide, Africa.
A honeyguide has thick skin to help protect it against bee stings.

Budgerigar

In Australia, you can see flocks of green-and-yellow birds zooming around the sky. These are budgerigars — small parrots with a long tail. It is hot and dry in their native habitat, and they fly great distances to find water and seeds. When it finally rains, huge numbers of budgerigars arrive from far and wide. The rain makes plants grow, which means there will be plenty for them to eat.

Budgerigars, or "budgies" for short, are some of the world's most popular pets. They are bred in many bright colours, including pale blue and purple. Their usual call is a cheerful chirrup, but budgies often learn to talk by copying the humans they hear.

Flocks of budgerigars can be huge, and contain thousands of birds.

Budgerigar, Australia.
Budgerigars nest in hollow trees. Several pairs may share the same tree.

A hibernating poorwill is almost as cold and still as a stone.

Common poorwill

Some creatures sleep the winter away until warmer weather returns. This deep sleep is called hibernation. Bears, bats, and many other animals do it, but only one bird can: the common poorwill. The poorwill can hibernate for many months. When winter comes, it hides among some rocks and lowers its body temperature to just 5°C (41°F). It slows its heartbeat and breathing until it hardly shows any signs of life. You can see why the Hopi people from North America call it "the sleeping one".

The poorwill is a nocturnal bird that flits through the dark sky like a huge moth, hunting insects. By day, it rests on the ground. It is named for the sound of its strange call, which sounds like "poor-will".

Common poorwill, North America. Grey and brown feathers hide the poorwill on the ground.

Little bee-eater

Little bee-eaters love to snuggle together. When they land on a branch, they all shuffle along to sit side by side. Flocks of little bee-eaters are a common sight in Africa. Usually, they live near rivers and in grasslands. They dig nest burrows in sand banks.

As you can guess from their name, these birds eat bees, but they also catch many wasps, too. They glide, soar, and swoop to grab their prey, then hit it against a branch and squeeze it hard. This kills the insect and gets rid of its sting, so it is safe to swallow. On average, they catch a bee or wasp every five minutes.

Little bee-eater, Africa.
Groups of bee-eaters line up in a row, all facing the same way.

At night, bee-eaters sleep in huddles to keep warm.

Common swift

The legs and feet of swifts are so tiny, you can barely see them.

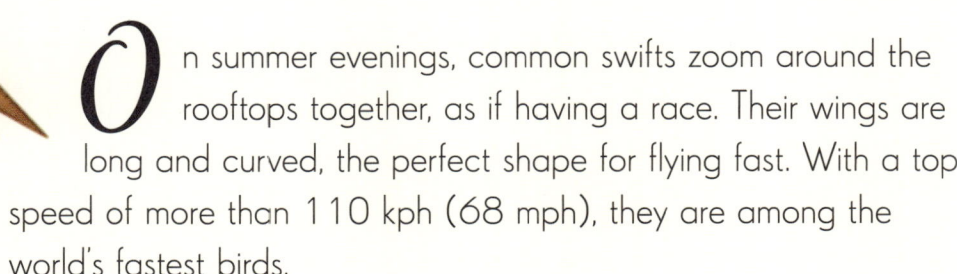

Common swift, Africa, Asia, and Europe. These masters of the sky can stay aloft for 10 months at a time.

On summer evenings, common swifts zoom around the rooftops together, as if having a race. Their wings are long and curved, the perfect shape for flying fast. With a top speed of more than 110 kph (68 mph), they are among the world's fastest birds.

Swifts catch insects in the air by flying with their mouths open. They also drink and bathe in flight, by dipping to the water. And they even sleep in the air! We think they probably climb higher and higher, then take short naps while gliding in circles. But the swifts must land to breed. Mostly, they nest on buildings, under the roof, or in nest boxes provided by people. After summer in Europe and Asia, they fly south to Africa, and return north in spring.

Southern cassowary

If you look at this pair of feet, you might mistake them for those of a Tyrannosaurus, but they belong to the southern cassowary. This large bird can't fly, but it can defend itself from predators with a lethal kick.

The long inner claw on a cassowary's foot is a sharp weapon

Willow ptarmigan

Willow ptarmigans live in cold northern regions where it snows heavily in winter. Unlike many other birds, they have feathers covering their legs and feet. The feathers help to keep their legs warm and turn their feet into snowshoes, which stop the birds sinking into the snow.

African fish eagle

Birds of prey have feet with large toes and sharp, curved claws called talons. This type of foot is designed to grab and hold prey. The talons of the African fish eagle are extra long for gripping slippery fish.

Sanderling

Most birds have four toes, three of which face forwards and one of which points backwards, but not the sanderling. This bird has lost its backwards-facing toe as it spends its life running around the seashore, and doesn't need to grip branches.

Mallard

Ducks, such as the mallard, spend their lives paddling in water. They have webbing between their toes that make their feet into flippers, like those used by divers. The large surface area helps them push against the water.

Ducks have webbed feet to help them swim.

Sulphur-crested cockatoo

Parrots, including cockatoos, spend a lot of time walking along branches. Their feet are designed for gripping — their two middle toes point forwards and the outer two point backwards, which gives them a pincer-like hold. These feet are also perfect for grasping food.

Wattled jacana

The prize for the bird with the longest toes might go to the wattled jacana. This bird lives in wetlands where it walks on top of floating water plants, such as water lilies, so needs wide, splayed toes to keep itself steady.

Eurasian treecreeper

Treecreepers are mouse-like birds that scurry over trees looking for insects to eat. They have thin toes but very long, curved claws to help them grip the bark. Their hold is so tight that they can even walk on the underside of branches.

Feet

The feet of birds may remind you of the scaly, clawed feet of their ancestors, the dinosaurs. Since birds don't have hands, they often use their feet to grasp objects or even scratch an itch. As with wings and beaks, the shape of a bird's feet can tell you a lot about where it lives. Sharp talons point to a bird of prey that needs to grip its food, while webbed feet indicate a life spent paddling on the water. Some birds even use their feet to show off to mates!

Wilson's bird of paradise

People once thought that birds of paradise lived in the sky and floated among the clouds. This myth is how these spectacular birds got their name, but actually they live in rainforests. The male Wilson's bird of paradise is one of the most colourful birds on Earth. The reason for his fancy dress is, of course, to attract females. But to truly show off his feathers, he first does some cleaning. He sweeps leaves off a patch of forest floor, which gives the females the best possible view of his display. Now, he perches in the middle of his "stage" and twirls two long tail feathers to make them shine. At the same time, he shouts his mating call, which sounds like a car alarm.

Male   Female

Wilson's bird of paradise, Indonesia.
The male of this species has a bare blue head, and incredibly curly tail feathers.

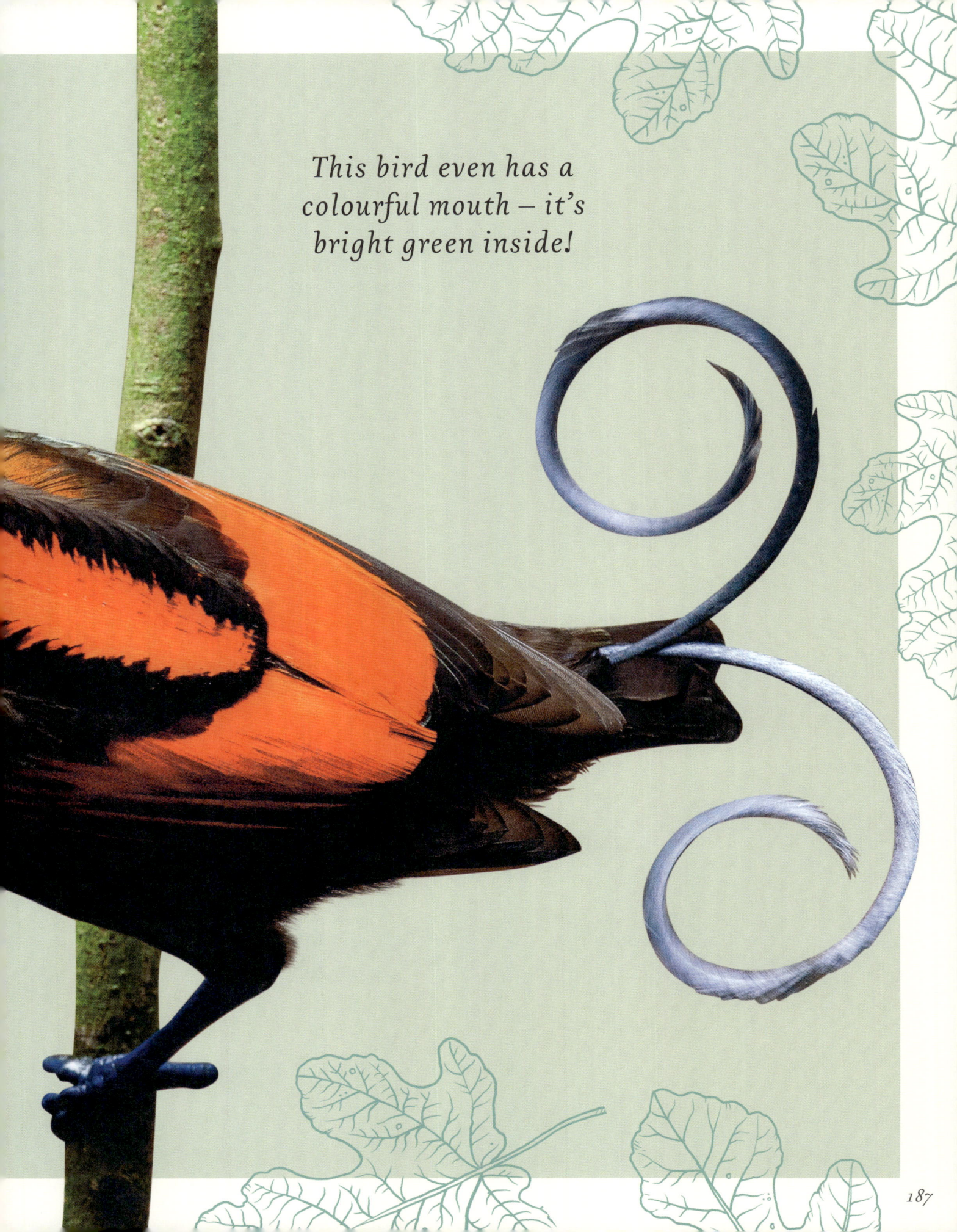

This bird even has a colourful mouth – it's bright green inside!

Male Female

House sparrow

House sparrow, worldwide, except Antarctica. Adult house sparrows continue to care for young sparrows that have recently left the nest.

Thousands of years ago, house sparrows lived only in Europe and western Asia. But they like living near humans so much, they have spread around the globe. Today, they thrive on every continent apart from Antarctica. These cheerful little birds nest on our buildings, and find food in fields and other areas around our cities, towns, and villages. House sparrows have even been spotted on the 80th floor of a skyscraper in New York, USA!

Male sparrows have a black throat patch. This acts like a badge that shows females how fit and strong they are. The more black there is, the healthier the male is. Sparrows often roll around in dirt. Believe it or not, one of these "dust baths" actually helps them to keep their feathers clean!

The chirp and cheep of house sparrows is one of the world's most familiar bird sounds.

Woodpecker finch, Galápagos Islands. The finch probes a branch with its twig tool.

Woodpecker finch

Far away from the mainland, the Galápagos Islands are home to many special animals that live nowhere else in the world. One of them is the woodpecker finch. It has an incredible skill. The finch picks up a twig, snaps it into a short piece, and holds it in its beak like a spear. Now the clever bird pokes its "spear" into holes in branches. The finch is using the twig as a tool to find food! With the help of the twig, it can extract juicy insect grubs from the wood that it couldn't otherwise reach.

We use tools every day, but not many birds do. Other birds with this rare skill include parrots and crows.

The woodpecker finch sometimes pulls a spine off a cactus to use as a tool.

Sociable weaver

The world's biggest bird nests are not made by big birds. They belong to little birds the size of sparrows — sociable weavers. Their stick nests can be larger than a garden shed and weigh as much as a cow! Many pairs of weavers team up to build their shared nest, usually on a tree or telegraph pole. If one bird doesn't do much work, the others soon tell it off. The nest's thick walls are good protection against predators and the hot sun.

Inside the structure, every breeding pair of weavers has their own nest chamber. Cheekily, all sorts of other birds, such as falcons and owls, may also move in. It can get quite crowded in the supersize nest.

Some weaver nests are around 100 years old.

Sociable weaver, Africa. Each nest may be home to several hundred pairs of weavers. Every pair enter their part of the nest through their own entrance hole.

Female long-tailed tits lay up to 12 speckled eggs.

Long-tailed tit

These pretty little birds have a round body with a very long tail, so they're a similar shape to lollipops! They are always on the move, as if in a hurry to get somewhere. You can see them flitting through woods and gardens like little acrobats. They search under every twig for caterpillars, spiders, and other small prey.

In early spring, the male and female make an astonishing nest from moss and spiders' webs. It looks like a shaggy coconut, with a domed roof and an entrance on one side. The pair line their cosy nest with hundreds of feathers to keep the chicks warm. Raising them is hard work, but the parents often have help from several of their siblings who share their territory. These helpers are the uncles and aunts of the chicks.

Long-tailed tit, Asia and Europe.
Long-tailed tit nests are stretchy, so can expand to fit around the growing chicks.

European robins are a symbol of Christmas in several countries.

European robin

Many small birds around the world are called "robins", but though they share a name, they are not related. European robins are a common sight in woods and gardens, where they can be amazingly tame. Often they follow gardeners, to eat any worms that are revealed by digging.

European robins sing all year, even in the middle of winter. Strangely, in towns and cities, you can hear them sing at night. Perhaps they sing in the dark because it is quieter then, or maybe the bright streetlights make them think it's daytime. Robins are famous for building nests in peculiar places. Some have been known to nest in coat pockets, postboxes, and even car bonnets!

European robin, Africa, Asia, and Europe. In cold weather, European robins fluff out their feathers to keep warm.

Elf owl, North America. Tall cacti give elf owls a safe, but spiky, home.

Elf owl

Elf owls are even smaller than sparrows! But these tiny hunting birds have a sharp beak and even sharper claws, so are pretty fierce for their size. They live in deserts and hunt at night, using their superb hearing and night vision to find prey in the dark. Mostly they catch small creatures such as moths, beetles, and spiders. They know how to tackle scorpions too, which have venomous stings, so can be dangerous. First, they bite off the stinger to disarm the scorpion, and then they swallow it whole.

These owls don't make nests of their own. Instead, they move into an old woodpecker hole already made in a tree or cactus. The parent owls are noisy birds and sound just like yapping puppies.

Elf owls are the smallest owls in the world.

ʻIʻiwi

West of North America, far out in the Pacific Ocean, lie the islands of Hawaii. They have many wonderful birds you won't see elsewhere. One is the ʻiʻiwi (you say its name "ee–EE–vee"). This small, red-and-black bird lives in wet forests. It perches on flowers and pushes its curved beak inside to lap up the nectar. There is one tree with frilly red flowers that it loves in particular. If it finds one of these trees, the bird guards it, to keep all the nectar for itself.

The ʻiʻiwi belongs to a family of birds called honeycreepers, found only in Hawaii. Each kind of honeycreeper has a different beak to suit what it eats. Some have chunky beaks, while others are pointed or curved.

The 'i'iwi sits out in the rain to bathe its feathers.

'I'iwi, Hawaii.
The curved beak of the 'i'iwi fits neatly into flowers.

Male
Female

Superb fairywren

Superb fairywrens are some of the best-loved birds in Australia. The male has a glittering blue head that sparkles when he moves. The female is mostly brown, but like him, has long legs and a long, perky tail. For some reason, the fairywrens always hold their tail upright. They live in bush country and gardens, hopping along the ground to hunt for insects.

Many scientists have studied superb fairywrens, because their family life is so interesting. When the male and female form a pair to breed, they have up to seven other fairywrens helping them. We have also discovered that these birds have around 20 different calls, all for different things. They even have different personalities. Some are bold, for example, while others are shy.

Black-backed kingfisher

A tiny bird whizzes past in a blur of colour. When it finally lands on a branch, you can fully admire its jewel-like feathers and bright red beak. This speedy bird is a black-backed kingfisher. It lives in tropical rainforests, usually beside streams. Like other kingfishers, it perches above the water and waits quietly until it sees a fish. Then, quick as a flash, it dives to snatch its prey. As well as fish, this kingfisher catches frogs and large insects in the rainforest. Sometimes, it even grabs insects already caught in spiders' webs!

There are many stories and myths about kingfishers. The ancient Greeks thought that these birds had magical powers over the sea, and were able to calm the waves, in order to nest on the surface.

Kingfishers make barely a splash as they dive into the water.

Black-backed kingfisher, Asia.
The kingfisher seizes fish and other prey in its dagger-like beak.

Common tailorbird, Asia. A tailorbird's snug nest is comfortably lined with cobwebs, grass, and soft plants.

Common tailorbird

Bird nests can be works of art. The common tailorbird, a busy little bird found in southern Asia, makes one of the most amazing nests of all. It is the female tailorbird which does all the work. She chooses a nice, strong leaf on a tree or bush. Next, she uses her feet to bend it round to form a cup shape. Now, she makes tiny holes around the edge of the leaf with her sharp beak. Finally, she uses her beak like a needle, to thread a strand of tree bark or spider silk through the holes. This stitches the leaf together. That's right — these birds can actually sew! When complete, the nest can hold up to five chicks safely.

The tailorbird uses around 200 stitches to make a nest.

Cuban tody, Cuba. Male and female todies look the same, so you can't tell which is which.

Todies are smaller than a tennis ball — and much more colourful!

Cuban tody

These tiny birds seem to have been painted with bright colours. No other birds have quite such an unusual combination of green, red, blue, and pink feathers. There are five kinds of tody, all of which live on islands in the Caribbean — Cuban todies are found only on the island of Cuba.

Despite their dazzling appearance, Cuban todies can be tricky to spot. They like to sit very still in the forests where they live, so are easy to miss. However, if an insect flies past, they dart from their perch to snap it up. These busy little birds feed each of their chicks up to 140 insects a day.

Hummingbirds

Violet sabrewing, Central America. Thanks to its curved beak, this species can feed from tube-shaped flowers.

Rufous-tailed hummingbird, Central America. Sugar water feeders in gardens often attract this hummingbird.

Bee hummingbird, Cuba. This tiny bird builds a nest the size of a bottle-top!

Anna's hummingbird, North America. The pink head and throat of the male Anna's hummingbird sparkle in the light.

White-booted racket-tail, northern South America. The male of this species shows off his long tail in his breeding display.

Tufted coquette, northern South America. The male tufted coquette has a bushy crest, and a magnificent fan of cheek feathers.

Rufous hummingbird, North America. This tiny species flies 1,500 km (930 miles) between its summer and winter homes.

Hummingbirds are the only birds able to fly backwards.

Sword-billed hummingbird, northwestern South America. What a beak! No other bird has a beak longer than its body, like this hummingbird does.

Hummingbirds are the fairies of the bird world. They're small, glittery, and flit about the forest. The tiny bee hummingbird is the smallest bird on Earth! As hummingbirds dart to and fro, it's hard to keep track of them. Their wings beat up to 200 times a second, which makes the loud humming sound after which they are named. To lead such active lives, they use a terrific amount of energy. So, all day long, they hover around flowers, sipping on sugar-rich nectar.

All hummingbirds live in North, Central, and South America. Although most are found in forests, some flock to gardens, where people hang up special feeders full of sugar water to attract them.

Penguins

Penguins are instantly recognizable from their plump bodies and black and white feathers. This group of birds has adapted to a life hunting under the sea. Unlike other birds, penguins have dense bones to help them dive and a very thick layer of stiff feathers with down underneath to keep them warm in water.

Hummingbirds

Hummingbirds beat their wings so fast, our eyes see them as a blur. The scientific name for the order to which both hummingbirds and swifts belong, Apodiformes, means "footless". Although not true – these birds do have little feet – they are usually seen flying.

Tree of life

There are many different types of bird, and we have barely scratched the surface of all their wondrous differences in this book. Scientists have named around 40 groups of birds, known as orders, which are further broken down into smaller groups, called families. This tree of life shows how closely these groups of birds are related.

Gamebirds

Including chickens, pheasants, peafowl, guineafowl, brush-turkeys and many more, the gamebirds are a diverse order. Many of these birds live on the ground and are weak fliers, but they can be marvellously colourful. Most eat a mix of plants and small animals.

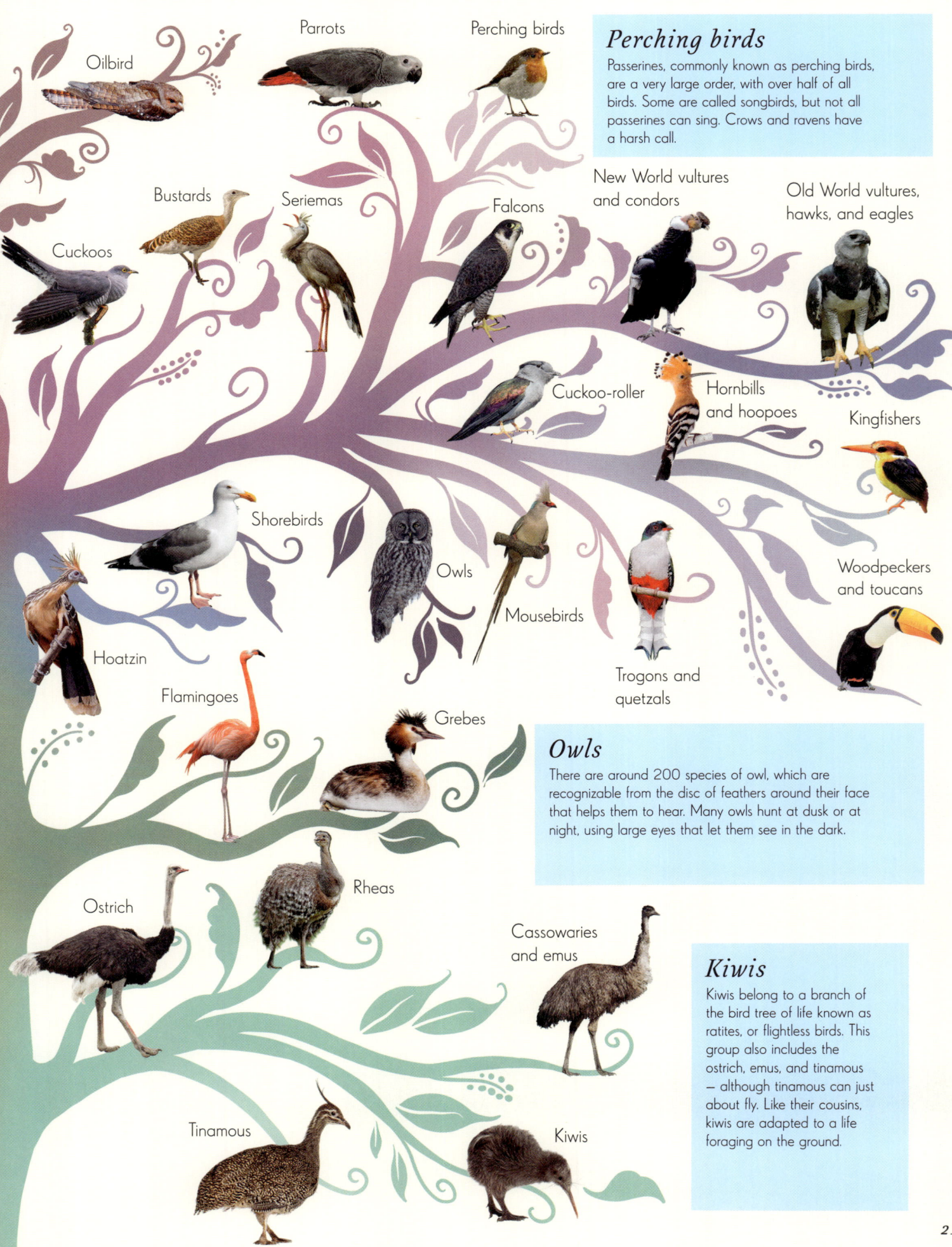

Glossary

amphibian Animal with a backbone that usually spends part of its life in water and the rest on land. It usually develops from an egg to a larva, and then into an adult. Frogs and newts are examples of amphibians

aquatic Description of an organism that lives in watery environments, such as rivers, lakes, and seas

beak Also known as a bill, the structure that makes up a bird's mouth. Beaks are used to perform many jobs, including collecting food and preening feathers. They are made up of plates of bone covered in the same tough material that makes up feathers

bill See "beak"

bird Animal with a backbone that has a hard beak and feathers. Most birds can fly and all lay hard-shelled eggs for which they often make a nest

breed Produce offspring

breeding season Time of year when animals produce young

camouflage Colour or pattern that disguise an animal where it lives, to help it hide from attackers

casque Structure found on the head of some birds, made from bone covered in the same tough material that makes up feathers

courtship display Dance performed by animals to attract a mate. During a display, a bird may show off its feathers or other ornaments, including brightly-coloured patches of skin, and call loudly

crest Tuft of feathers found on the head of some birds used to show off to mates

desert Habitat found where there is very little rainfall. Deserts are usually sandy, and can be hot or cold

domestic animal Animal that is bred to be kept by humans, either for meat, wool, and to do work, or as a pet

dinosaur Type of ancient reptile that dominated the Earth between around 243 and 66 million years ago. Birds evolved from one group of dinosaurs and share many of their features, such as feathers and the ability to lay hard-shelled eggs

down Type of feather that is very fluffy and used by birds to keep warm. Chicks only have down feathers

echolocation Using sound to work out how far away an object is by listening for the echo from a call. Some bats and a few birds, including oilbirds, use echolocation to find their way around

egg Capsule produced by animals to reproduce. Birds lay hard-shelled eggs in which their chicks develop. Bird eggs are usually round or oval-shaped and can have varied colours and patterns

endangered When an animal becomes very rare in the wild. If we do not do something to help, the animal might go extinct

evolution Process of a species changing over a long period of time, until it is so different a new species is created

extinct When the last member of a species dies and there are no more of its kind left anywhere on Earth

feather Thin, flat structure that is made from the same material as hair and which is unique to birds. Feathers cover a bird's body and help to keep it warm and waterproof. They also help flying birds take to the air and can be brightly coloured

fish Animal with a backbone that usually spends its whole life in water and breathes using gills

flipper Thin, flat limb that acts like a paddle to push an animal through water. Penguins' wings have become flippers

fossil Hardened remains of organisms that lived millions of years ago. Fossils can be of body parts, such as bones, or things made by life forms, such as footprints

habitat Place where animals, plants, and other living things are found. Habitats can be on land, or in water. Many species live only in particular types of habitat

herbivore Animal that eats only plants

hibernation Deep sleep that some animals fall into over winter. It can last for many months

incubation Process of keeping an egg warm so the baby animal inside can grow. Birds often build nests in which to incubate their eggs and sit on them to keep them warm

insect Invertebrate with three pairs of legs and a body in three sections. Many insects have wings

invertebrate Animals with no backbone, such as insects, spiders, and crabs

iridescence Colourful shine of light seen on some feathers, created by the structure of the feather

mammals Animal with a backbone that has warm blood and which has fur or hair. Most female mammals give birth to live babies and all feed them milk

mate Partner that an animal produces offspring with

marsh Habitat with low-growing plants and wet ground

migration Long journey made by animals to find a new place to feed or raise their families. Many animals migrate every year between their summer and winter homes

nectar Sweet, sugary liquid made by flowers. Insects and birds visit flowers to drink nectar

nest Structure built by some birds in which to lay their eggs. Nests can be constructed from twigs, mud, spider webs, moss, lichen, leaves, roots, and even a bird's saliva. Most birds build a new nest for each breeding season

nocturnal Description of an animal that is active at night

parasite Animal that lives on another animal, or inside its body, and causes the animal harm. It feeds on the animal and cannot live without it. Some cuckoos are known as brood parasites as they trick other birds into feeding their young for them

perch When birds stand on a branch or other surface and their toes lock around it to keep them steady. Also the place where they are standing

pigment Substance that gives something colour

pollination Process of pollen being moved between flowers, allowing the plant to make seeds. Pollen is often moved by animals, called pollinators, such as birds

predator Animal that hunts another animal, called prey, for food

preen Process of a bird running its beak through its feathers to make sure the surface lies flat, and sometimes to cover the feathers with a waterproof oil

prehistoric From a very long time ago. Many prehistoric animals and plants no longer exist, but we know about them from fossils

prey Animal that is hunted by a predator

rainforest Forest habitat where it is very wet and rains a lot

reptile Animal with a backbone that has tough skin, is covered in hard scales, and usually lays eggs. Reptiles include snakes, lizards, and turtles

scale Hard, flat plate found on the skin of some animals. Scales usually cover the legs and feet of birds

species Particular type of animal, plant, or other living thing. For example, ostriches and sunbitterns are different species of bird. Members of the same species can breed together to produce young, but they usually do not breed with other species

talons Long, sharp, curved claws that some predators use to kill their prey. Birds with talons include owls, eagles, hawks, and vultures

webbed Having skin between the toes. Many water birds have webbed feet to help them swim

wetland Area of low-growing plants that is very wet or constantly flooded

wing Arm of a bird that is covered in feathers. Wings come in different shapes depending on how they are used, for example, long, thin wings are used for soaring in the air

ultraviolet light Type of light invisible to us, but which can be seen by some other animals, including birds. Many birds have ultraviolet patterns on their feathers

Visual guide

Green peafowl, page 4
Pavo muticus
Location: Southeast Asia
Bird order: Gamebirds
Length: 3 m (10 ft)

Dalmatian pelican, page 6
Pelecanus crispus
Location: Asia and Europe
Bird order: Pelicans and relatives
Length: 1.8 m (6 ft)

Emu, page 8
Dromaius novaehollandiae
Location: Australia
Bird order: Emus and cassowaries
Height: 1.7 m (6 ft)

Marabou stork, page 10
Leptoptilos crumenifer
Location: Africa
Bird order: Storks
Height: 1.5 m (5 ft)

Secretary bird, page 12
Sagittarius serpentarius
Location: Africa
Bird order: Hawks and relatives
Height: 1.5 m (5 ft)

Red-crowned crane, page 14
Grus japonensis
Location: Eastern Asia
Bird order: Cranes and relatives
Height: 1.5 m (5 ft)

Black swan, page 16
Cygnus atratus
Location: Australia
Bird order: Waterfowl
Length: 1.4 m (5 ft)

Wandering albatross, page 18
Diomedea exulans
Location: Atlantic, Pacific, and Southern Oceans
Bird order: Petrels
Length: 1.4 m (5 ft)

Andean condor, page 20
Vultur gryphus
Location: South America
Bird order: New World vultures
Length: 1.2 m (4 ft)

Bearded vulture, page 22
Gypaetus barbatus
Location: Africa, Asia, and Europe
Bird order: Birds of prey
Length: 1.2 m (4 ft)

Harpy eagle, page 26
Harpia harpyja
Location: Central and South America
Bird order: Birds of prey
Length: 1.05 m (3 ft)

Great bustard, page 28
Otis tarda
Location: Asia and Europe
Bird order: Bustards
Length: 1.05 m (3 ft)

Red-tailed tropicbird, page 30
Phaethon rubricauda
Location: Indian and Pacific oceans
Bird order: Tropicbirds
Length: 1.05 m (3 ft)

Hyacinth macaw, page 32
Anodorhynchus hyacinthinus
Location: South America
Bird order: Parrots
Length: 1 m (3 ft)

Great frigatebird, page 34
Fregata minor
Location: Tropical oceans
Bird order: Cormorants and relatives
Length: 1 m (3 ft)

Lesser flamingo, page 36
Phoeniconaias minor
Location: Africa and Asia
Bird order: Flamingoes
Length: 1 m (3 ft)

Superb lyrebird, page 40
Menura novaehollandiae
Location: Australia
Bird order: Perching birds
Length: 1 m (3 ft)

Great cormorant, page 42
Phalacrocorax carbo
Location: Africa, Asia, Europe, North America, and Oceania
Bird order: Cormorants and relatives
Length: 1 m (3 ft)

Great northern diver, page 44
Gavia immer
Location: Europe and North America
Bird order: Divers
Length: 91 cm (36 in)

Red-legged seriema, page 46
Cariama cristata
Location: South America
Bird order: Seriemas
Length: 90 cm (35 in)

Rhinoceros hornbill, page 48
Buceros rhinoceros
Location: Southeast Asia
Bird order: Hornbills and hoopoes
Length: 90 cm (35 in)

Ring-necked pheasant, page 50
Phasianus colchicus
Location: Asia, Europe, and North America
Bird order: Gamebirds
Length: 89 cm (35 in)

Anhinga, page 52
Anhinga anhinga
Location: North and South America
Bird order: Cormorants and relatives
Length: 89 cm (35 in)

Roseate spoonbill, page 54
Platalea ajaja
Location: North, Central, and South America
Bird order: Pelicans and relatives
Length: 85 cm (33 in)

Blue-footed booby, page 56
Sula nebouxii
Location: North and South America
Bird order: Cormorants and relatives
Length: 84 cm (33 in)

Snow goose, page 58
Anser caerulescens
Location: North America
Bird order: Waterfowl
Length: 83 cm (33 in)

Great bittern, page 60
Botaurus stellaris
Location: Africa, Asia, and Europe
Bird order: Pelicans and relatives
Length: 80 cm (31 in)

Red junglefowl, page 62
Gallus gallus
Location: Asia
Bird order: Gamebirds
Length: 78 cm (31 in)

Australian brushturkey, page 64
Alectura lathami
Location: Australia
Bird order: Gamebirds
Length: 70 cm (28 in)

Scarlet ibis, page 66
Eudocimus ruber
Location: South America
Bird order: Pelicans and relatives
Length: 70 cm (28 in)

Great grey owl, page 68
Strix nebulosa
Location: Asia, Europe, and North America
Bird order: Owls
Length: 69 cm (27 in)

Common raven, page 72
Corvus corax
Location: Africa, Asia, Europe, and North America
Bird order: Perching birds
Length: 69 cm (27 in)

European herring gull, page 74
Larus argentatus
Location: Europe
Bird order: Shorebirds
Length: 67 cm (26 in)

Hoatzin, page 76
Opisthocomus hoazin
Location: South America
Bird order: Hoatzins
Length: 66 cm (26 cm)

Resplendent quetzal, page 78
Pharomachrus mocinno
Location: Central America
Bird order: Trogons and quetzals
Length: 65 cm (26 in)

Kākāpō, page 80
Strigops habroptila
Location: New Zealand
Bird order: Parrots
Length: 64 cm (25 in)

Helmeted guineafowl, page 82
Numida meleagris
Location: Africa
Bird order: Gamebirds
Length: 63 cm (25 in)

Toco toucan, page 84
Ramphastos toco
Location: South America
Bird order: Woodpeckers and toucans
Length: 61 cm (24 in)

Greater roadrunner, page 86
Geococcyx californianus
Location: North America
Bird order: Cuckoos
Length: 56 cm (22 in)

Harlequin duck, page 88
Histrionicus histrionicus
Location: Asia, Europe, and North America
Bird order: Waterfowl
Length: 54 cm (21 in)

Great crested grebe, page 90
Podiceps cristatus
Location: Africa, Asia, Europe, and Oceania
Bird order: Grebes
Length: 51 cm (20 in)

Cuckoo-roller, page 92
Leptosomus discolor
Location: Madagascar
Bird order: Cuckoo-rollers
Length: 50 cm (20 in)

Montezuma oropendola, page 94
Psarocolius montezuma
Location: Mexico and Central America
Bird order: Perching birds
Length: 50 cm (20 in)

Oilbird, page 98
Steatornis caripensis
Location: South America
Bird order: Oilbirds
Length: 50 cm (20 in)

Sunbittern, page 100
Eurypyga helias
Location: Central and South America
Bird order: Sunbitterns
Length: 48 cm (19 in)

Green heron, page 102
Butorides virescens
Location: North and Central America
Bird order: Pelicans and relatives
Length: 48 cm (19 in)

Laughing kookaburra, page 104
Dacelo novaeguineae
Location: Australia
Bird order: Kingfishers and relatives
Length: 47 cm (19 in)

Northern potoo, page 106
Nyctibius jamaicensis
Location: The Caribbean and Central America
Bird order: Potoos
Length: 46 cm (18 in)

Pileated woodpecker, page 108
Dryocopus pileatus
Location: North America
Bird order: Woodpeckers
Length: 46 cm (18 in)

Little spotted kiwi, page 110
Apteryx owenii
Location: New Zealand
Bird order: Kiwis
Length: 45 cm (18 in)

Little penguin, page 112
Eudyptula minor
Location: Australia and New Zealand
Bird order: Penguins
Length: 45 cm (18 in)

Peregrine falcon, page 114
Falco peregrinus
Location: Worldwide, except Antarctica
Bird order: Falcons
Length: 45 cm (18 in)

Cape sugarbird, page 116
Promerops cafer
Location: South Africa
Bird order: Perching birds
Length: 43 cm (17 in)

Livingstone's turaco, page 118
Tauraco livingstonii
Location: Africa
Bird order: Turacos
Length: 43 cm (17 in)

Bar-tailed godwit, page 120
Limosa lapponica
Location: Africa, Asia, Europe, North America, and Oceania
Bird order: Shorebirds
Length: 41 cm (16 in)

Snow petrel, page 122
Pagodroma nivea
Location: Antarctica and Southern Ocean
Bird order: Petrels
Length: 40 cm (16 in)

Pink pigeon, page 124
Nesoenas mayeri
Location: Mauritius
Bird order: Pigeons
Length: 40 cm (16 in)

Grey parrot, page 126
Psittacus erithacus
Location: Africa
Bird order: Parrots
Length: 39 cm (15 in)

Tufted puffin, page 128
Fratercula cirrhata
Location: Asia and North America
Bird order: Shorebirds
Length: 38 cm (15 in)

Scissor-tailed flycatcher, page 130
Tyrannus forficatus
Location: North and Central America
Bird order: Perching birds
Length: 38 cm (15 in)

Eurasian coot, page 132
Fulica atra
Location: Asia, Africa, Europe, and Oceania
Bird order: Cranes and relatives
Length: 38 cm (15 in)

Arctic tern, page 134
Sterna paradisaea
Location: Oceans worldwide
Bird order: Shorebirds
Length: 36 cm (14 in)

Common cuckoo, page 136
Cuculus canorus
Location: Africa, Asia, and Europe
Bird order: Cuckoos
Length: 33 cm (13 in)

White tern, page 140
Gygis alba
Location: Indian and Pacific Oceans
Bird order: Shorebirds
Length: 33 cm (13 in)

Ruff, page 142
Calidris pugnax
Location: Africa, Asia, and Europe
Bird order: Shorebirds
Length: 32 cm (13 in)

Eurasian hoopoe, page 144
Upupa epops
Location: Africa, Asia, and Europe
Bird order: Hornbills and hoopes
Length: 32 cm (13 in)

Andean cock-of-the-rock, page 146
Rupicola peruvianus
Location: South America
Bird order: Perching birds
Length: 32 cm (13 in)

Blue jay, page 148
Cyanocitta cristata
Location: North America
Bird order: Perching birds
Length: 30 cm (12 in)

Bearded bellbird, page 150
Procnias averano
Location: South America
Bird order: Perching birds
Length: 28 cm (11 in)

Namaqua sandgrouse, page 152
Pterocles namaqua
Location: Africa
Bird order: Sandgrouses
Length: 28 cm (11 in)

Cuban trogon, page 154
Priotelus temnurus
Location: Cuba
Bird order: Trogons and quetzals
Length: 28 cm (11 in)

Killdeer, page 156
Charadrius vociferus
Location: North, Central, and South America
Bird order: Shorebirds
Length: 26 cm (10 in)

Flame bowerbird, page 158
Sericulus ardens
Location: New Guinea
Bird order: Perching birds
Length: 25 cm (10 in)

Long-tailed manakin, page 160
Chiroxiphia linearis
Location: Central America
Bird order: Perching birds
Length: 23 cm (9 in)

European starling, page 162
Sturnus vulgaris
Location: Africa, Asia, and Europe
Bird order: Perching birds
Length: 22 cm (9 in)

Northern cardinal, page 166
Cardinalis cardinalis
Location: North America
Bird order: Perching birds
Length: 22 cm (9 in)

Noisy pitta, page 168
Pitta versicolor
Location: Oceania
Bird order: Perching birds
Length: 21 cm (8 in)

White-throated dipper, page 170
Cinclus cinclus
Location: Asia and Europe
Bird order: Perching birds
Length: 21 cm (8 in)

Bohemian waxwing, page 172
Bombycilla garrulus
Location: Asia, Europe, and North America
Bird order: Perching birds
Length: 21 cm (8 in)

Greater honeyguide, page 174
Indicator indicator
Location: Africa
Bird order: Woodpeckers
Length: 20 cm (8 in)

Budgerigar, page 176
Melopsittacus undulatus
Location: Australia
Bird order: Parrots
Length: 20 cm (8 in)

Common poorwill, page 178
Phalaenoptilus nuttallii
Location: North America
Bird order: Nightjars
Length: 20 cm (8 in)

Little bee-eater, page 180
Merops pusillus
Location: Africa
Bird order: Kingfishers and relatives
Length: 17 cm (7 in)

Common swift, page 182
Apus apus
Location: Africa, Asia, and Europe
Bird order: Hummingbirds and swifts
Length: 17 cm (7 in)

Wilson's bird of paradise, page 186
Diphyllodes respublica
Location: Indonesia
Bird order: Perching birds
Length: 16 cm (6 in)

House sparrow, page 188
Passer domesticus
Location: Worldwide, except Antarctica
Bird order: Perching birds
Length: 16 cm (6 in)

Woodpecker finch, page 190
Camarhyncus pallidus
Location: Galápagos Islands
Bird order: Perching birds
Length: 15 cm (6 in)

Sociable weaver, page 192
Philetairus socius
Location: Africa
Bird order: Perching birds
Length: 14 cm (6 in)

Long-tailed tit, page 194
Aegithalos caudatus
Location: Asia and Europe
Bird order: Perching birds
Length: 14 cm (6 in)

European robin, page 196
Erithacus rubecula
Location: Africa, Asia, and Europe
Bird order: Perching birds
Length: 14 cm (6 in)

Elf owl, page 198
Micrathene whitneyi
Location: North America
Bird order: Owls
Length: 14 cm (6 in)

'I'iwi, page 200
Drepanis coccinea
Location: Hawaii
Bird order: Perching birds
Length: 14 cm (6 in)

Superb fairywren, page 202
Marulus cyaneus
Location: Australia
Bird order: Perching birds
Length: 14 cm (6 in)

Black-backed kingfisher, page 204
Ceyx erithaca
Location: Asia
Bird order: Kingfishers and relatives
Length: 14 cm (6 in)

Common tailorbird, page 206
Orthotomus sutorius
Location: Asia
Bird order: Perching birds
Length: 13 cm (5 in)

Cuban tody, page 208
Todus multicolor
Location: Cuba
Bird order: Kingfishers and relatives
Length: 11 cm (4 in)

Bee hummingbird, page 210
Mellisuga helenae
Location: Cuba
Bird order: Hummingbirds and swifts
Length: 6 cm (2 in)

 **Penguin Random House**

Senior editor Olivia Stanford
Designers Charlotte Jennings, Roohi Rais, Ann Cannings
Editor Abi Maxwell
Assistant picture researcher Nunhoih Guite
DTP designer Dheeraj Singh
Managing editor Gemma Farr
Managing art editors Elle Ward, Ivy Sengupta
Jacket coordinator Elin Woosnam
Senior production editor Nikoleta Parasaki
Senior production controller Ben Radley
Delhi creative head Malavika Talukder
Managing director Sarah Larter

Consultant Sacha Barbato

First published in Great Britain in 2024 by
Dorling Kindersley Limited
DK, 20 Vauxhall Bridge Road,
London, SW1V 2SA

The authorised representative in the EEA is
Dorling Kindersley Verlag GmbH. Arnulfstr. 124,
80636 Munich, Germany

Text copyright © Ben Hoare 2024
Copyright © 2024 Dorling Kindersley Limited
A Penguin Random House Company
10 9 8 7 6 5 4 3
013–340978–Sept/2024

All rights reserved.
No part of this publication may be reproduced, stored in or introduced into a retrieval system, or transmitted, in any form or by any means (electronic, mechanical, photocopying, recording, or otherwise), without the prior written permission of the copyright owner.

A CIP catalogue record for this book
is available from the British Library.
ISBN: 978-0-2416-7496-3

Printed and bound in China

www.dk.com

This book was made with Forest Stewardship Council™ certified paper - one small step in DK's commitment to a sustainable future. **For more information go to www.dk.com/our-green-pledge**

DK would like to thank: Sonny Flynn and Brandie Tully-Scott for additional design; Lois Ware for proofreading; Daniel Long for the feature illustrations; Angela Rizza for the pattern and cover illustrations; and Tuba Syed for metatagging.

About the author: Ben Hoare has been fascinated by wildlife since he was a toddler. He was the features editor of BBC Wildlife Magazine, and has been an editor and writer of many DK books, including *An Anthology of Intriguing Animals*, and *The Wonders of Nature*.

Picture credits

The publisher would like to thank the following for their kind permission to reproduce their photographs:
(Key: a-above; b-below/bottom; c-centre; f-far; l-left; r-right; t-top)

4-5 123RF.com: iamtk. **6-7 Shutterstock.com:** JaklZdenek. **8-9 Shutterstock.com:** Cassandra Cury. **10-11 naturepl.com:** Sylvain Cordier. **12-13 123RF.com:** Johanswan. **14-15 Alamy Stock Photo:** Papilio / Robert Pickett. **16 Alamy Stock Photo:** FLPA. **19 Shutterstock.com:** Agami Photo Agency. **20-21 Thomas Fuhrmann:** www.snowmanstudios.de. **23 Depositphotos Inc:** Slowmotiongli. **24 Dorling Kindersley:** Peter Minister, Digital Sculptor (cr). **25 Alamy Stock Photo:** Clarence Holmes Wildlife (br); Nature Photographers Ltd / Paul R. Sterry (crb). **Dorling Kindersley:** Jon Hughes (ca). **Dreamstime.com:** Dragoneye (bc). **Phillip Krzeminski:** (c). **26-27 Alamy Stock Photo:** Teila K. Day Photography. **28-29 Dreamstime.com:** Volodymyr Byrdyak. **30-31 Jacob Drucker. 32-33 Alamy Stock Photo:** Nature Picture Library / Bence Mate. **34-35 Alamy Stock Photo:** DanitaDelimont / Yuri Choufour. **36 Shutterstock.com:** Bruce Seabrook. **38 Alamy Stock Photo:** Mats Janson (bl). **Dreamstime.com:** Agami Photo Agency (tl); Jim Cumming (cl). **38-39 Alamy Stock Photo:** Monkey Business (tc); Nature Photographers Ltd / Paul R. Sterry (bc). **39 Alamy Stock Photo:** Blickwinkel / McPHOTO / MAS (cla); Kevin Elsby (bl); FLPA (br). **Dreamstime.com:** Dwiputra18 (crb). **Shutterstock.com:** Robert Harding Video (tr). **40-41 Alamy Stock Photo:** Minden Pictures. **43 Alamy Stock Photo:** Nature Picture Library / Markus Varesvuo. **44 Alamy Stock Photo:** All Canada Photos / Roberta Olenick. **46 Kacau Oliviera / Solent News & Photo Agency. 49 naturepl.com:** Tim Laman. **50 Depositphotos Inc:** JakubMrocek (t). **Shutterstock.com:** Wang LiQiang (bl); Wang LiQiang (br). **51 Dreamstime.com:** Diego Grandi (t). **Shutterstock.com:** Sunti (b). **52-53 Sarathlal Sasidharan. 54-55 Dreamstime.com:** Isselee. **56-57 Alamy Stock Photo:** Minden Pictures / Ingo Arndt. **58-59 naturepl.com:** Jack Dykinga. **60-61 Alamy Stock Photo:** Minden Pictures / Cees Uri / NIS. **62-63 Dreamstime.com:** Foto Journey. **65 naturepl.com:** BIA / Jan Wegener. **66-67 Alamy Stock Photo:** Nature Picture Library / Sylvain Cordier. **68-69 Shutterstock.com:** Jim Cumming. **70 naturepl.com:** Charlie Hamilton James (ca); Photo Ark / Joel Sartore (cb); Michael D. Kern (br). **Science Photo Library:** Natural History Museum, London (c). **70-71 naturepl.com:** Yves Lanceau (tc); Staffan Widstrand (cc). **71 Alamy Stock Photo:** imageBROKER / Phil McLean (bl). **naturepl.com:** Michael Durham (cr). **Science Photo Library:** Mark Sykes (br); Dr Keith Wheeler (tc). **72-73 naturepl.com:** Niall Benvie. **74-75 Science Photo Library:** Chris Hellier. **76-77 Alamy Stock Photo:** Minden Pictures / Nate Chappell / BIA. **78-79 Getty Images:** 500px / Daniel Parent. **80 Sam O'Leary. 82-83 Alamy Stock Photo:** Blickwinkel / McPHOTO / MAS. **84-85 Shutterstock.com:** Eric Isselee. **86-87 Getty Images:** Moment / Jeff R Clow. **88-89 naturepl.com:** Gerrit Vyn. **90-91 Alamy Stock Photo:** Mark Sandbach. **92 ©Paul van Giersbergen. 93 Science Photo Library:** Tony Camacho. **94-95 Alamy Stock Photo:** Minden Pictures / Era-Images / Colin Harris (br); Minden Pictures / Konrad Wothe (tc); imageBROKER / Frank Derer (tr); William Leaman (bl). **Shutterstock.com:** Danita Delimont (cl). **97 Alamy Stock Photo:** imageBROKER / Neil Bowman (tr); Minden Pictures / Ingo Arndt (tl); Dominic Robinson (bl); Cro Magnon (br). **Shutterstock.com:** Feng Yu (cb). **98-99 Shutterstock.com:** Traveller MG. **100-101 Getty Images / iStock:** GlobalP. **102-103 123RF.com:** Gonepaddling. **104-105 Alamy Stock Photo:** Christian Htter. **106-107 Shutterstock.com:** Fabio Maffei. **108 Alamy Stock Photo:** Minden Pictures / Donald M. Jones. **110-111 Alamy Stock Photo:** Nature Picture Library / Tui De Roy. **112-113 Alamy Stock Photo:** Auscape International Pty Ltd / Ian Beattie / Auscape. **114-115 naturepl.com:** Luke Massey. **116-117 Getty Images / iStock:** Neil Bowman. **119 naturepl.com:** Roland Seitre. **121 naturepl.com:** Markus Varesvuo. **122-123 Depositphotos Inc:** Tarpan. **124 Alamy Stock Photo:** Peter Schickert (b). **Minden Pictures:** Martin Withers (tr). **naturepl.com:** Ann & Steve Toon (tl). **125 Alamy Stock Photo:** Biosphoto / Jean-Francois Noblet (c); Minden Pictures / Hans Glader / BIA (bl); Minden Pictures / Greg Oakley / BIA (br). **Getty Images:** Moment / Rapeepong Puttakumwong (tr). **126-127 Depositphotos Inc:** Lifeonwhite. **128-129 © Christopher Dodds:** www.chrisdoddsphoto.com. **130-131 naturepl.com:** Alan Murphy. **132-133 Alamy Stock Photo:** Minden Pictures / Natalia Paklina / Buiten-beeld. **134-135 Dreamstime.com:** Hakoar. **136-137 naturepl.com:** Hermann Brehm. **138 Getty Images:** Corbis / Paul Starosta (tl); Corbis / Paul Starosta (tc); Corbis / Paul Starosta (bl); Corbis / Paul Starosta (crb). **139 Alamy Stock Photo:** Science History Images / Photo Researchers (cla). **Dreamstime.com:** Isselee (b). **Getty Images:** Corbis / Paul Starosta (t); Corbis / Paul Starosta (clb). **Science Photo Library:** DK Images (tr). **141 Alamy Stock Photo:** Oliver Smart. **142 Depositphotos Inc:** Davem1972 (b). **143 Alamy Stock Photo:** Minden Pictures / Winfried Wisniewski. **144-145 naturepl.com:** BIA / Thomas Hinsche. **146-147 naturepl.com:** Morley Read. **148-149 Alamy Stock Photo:** Chris Hennessy. **150 Douglas Greenberg. 152-153 Alamy Stock Photo:** Blickwinkel / M. Woike. **154-155 Alamy Stock Photo:** All Canada Photos / Glenn Bartley. **156 Alamy Stock Photo:** James Schaedig. **158 Getty Images / iStock:** Banu R. **160-161 Getty Images:** Moment / © Juan Carlos Vindas. **162-163 Dreamstime.com:** Agdbeukhof. **164 Alamy Stock Photo:** Blickwinkel / McPHOTO / MAS (cr); Jerome Murray - CC (tr); McPhoto / Rolfes (cla). **164-165 Alamy Stock Photo:** Raimund Linke (c). **165 Alamy Stock Photo:** AGAMI Photo Agency / Ralph Martin (tr); Steve Cushing (tc); David DesRochers (cra); Blickwinkel / Mcphoto / Mas (br). **166-167 Dreamstime.com:** Thomas Torget. **168-169 Alamy Stock Photo:** Minden Pictures / Eric Sohn Joo Tan / BIA. **170-171 naturepl.com:** Markus Varesvuo. **172 Alamy Stock Photo:** imageBROKER.com GmbH & Co. KG / D. Usher. **175 naturepl.com:** Roland Seitre. **176 naturepl.com:** Steven David Miller. **178-179 Alamy Stock Photo:** Rick & Nora Bowers. **180-181 Alamy Stock Photo:** Kit Day. **182-183 Shutterstock.com:** Dilomski. **184 Alamy Stock Photo:** AfriPics.com (bl); Trevor Collens (t); Panoramic Images (c). **Shutterstock.com:** Bildagentur Zoonar GmbH (crb). **185 Alamy Stock Photo:** Alan Spencer Norfolk (tc); Robertharding / G & M Therin-Weise (tr); Colin Varndell (br). **186-187 Alamy Stock Photo:** Minden Pictures / Ch'ien Lee. **188-189 naturepl.com:** Philippe Clement. **190 Alamy Stock Photo:** Minden Pictures / D. Parer & E. Parer-Cook. **192-193 Getty Images / iStock:** Wirestock. **194 Shutterstock.com:** Coulanges. **196-197 Shutterstock.com:** Kristian Bell. **198-199 Minden Pictures:** FLPA. **200-201 Shutterstock.com:** Kendall Collett. **203 Shutterstock.com:** Leonie Ailsa Puckeridge. **205 Getty Images / iStock:** Abdul Sameer. **206 Depositphotos Inc:** SyedFAbbas. **208-209 Shutterstock.com:** Milan Zygmunt. **210 Andy Morffew (tr,br). Shutterstock.com:** Keneva Photography (bl); Richard Winston (tl); Piotr Poznan (c). **211 Alamy Stock Photo:** All Canada Photos / Glenn Bartley (tr); Keith Allen (tl). **Getty Images / iStock:** Ken Canning (c). **212 123RF.com:** Keith Levit / keithlevit (cla); Mike Price / Mhprice (cr/Turacos). **Alamy Stock Photo:** Saverio Gatto (c); Peter Schickert (crb); Robertharding / James Hager (cb). **Depositphotos Inc:** Dianaarturovna (cr). **Dreamstime.com:** Inaras (ca/Penguins); Tarpan (ca); Alexander Potapov (crb/Waterfowl); Jacob Drucker (cra). **Minden Pictures:** Martin Withers (crb/Messites). **Shutterstock.com:** Foto Journey (crb/Gamebirds); Piotr Poznan (tr). **213 Alamy Stock Photo:** All Canada Photos / Glenn Bartley (c/Trogon); Gabbro (cra); VWPics / Jon G. Fuller (cra/Eagles); Wildlife / Robert McGouey (c). **Depositphotos Inc:** Lifeonwhite (tr); Panuruangjan (cra/Kingfisher). **Dreamstime.com:** Dragoneye (cb/Emu); Vasyl Helevachuk (cra/Hoopoe); Tupungato (cl/Herring gull); Sombra12 (cb); Igor Stramyk (clb); Rudolf Ernst (cb/Rheas); Isselee (bc); Isselee (bc/Kiwi). **Getty Images / iStock:** lvkuzmin (cl). **naturepl.com:** Hermann Brehm (cla/Cuckoos); Photo Ark / Joel Sartore (ca/Cuckoo roller). **Science Photo Library:** Tony Camacho (ca/bustard). **Shutterstock.com:** Kristian Bell (tc/Songbirds); Traveller MG (tl); Eric Isselee (cr); Eric Isselee (cb/crested grebe). **Kacau Oliviera / Solent News & Photo Agency:** (cla).

Cover images: Front: 123RF.com: Kajornyot tc; **Alamy Stock Photo:** All Canada Photos / Glenn Bartley cra, Kit Day clb, Minden Pictures / Thomas Marent cla, WILDLIFE GmbH cl; **Depositphotos Inc:** Lifeonwhite crb; **Dreamstime.com:** Cowboy54 br, Perchhead tr; **Getty Images / iStock:** Guenterguni cra/ (frigatebird); **naturepl.com:** Juergen & Christine Sohns bl

All other images © Dorling Kindersley Limited.